Charles Seale-Hayne Library
# University of Plymouth
**(01752) 588 588**
LibraryandITenquiries@plymouth.ac.uk

1995

# Tree-Ring Dating and Archaeology

CROOM HELM STUDIES IN ARCHAEOLOGY

General Editor: Leslie Alcock, University of Glasgow


SURVEYING FOR ARCHAEOLOGISTS AND OTHER FIELDWORKERS
A.H.A. Hogg

CELTIC CRAFTMANSHIP IN BRONZE
H.E. Kilbride-Jones

EARLY MAN IN BRITAIN AND IRELAND
Alex Morrison

THE PALAEOLITHIC AGE
J.J. Wymer

A rare signature pattern in samples from Trinity College, Dublin (Library), Coagh, Co. Tyrone, and Hillsborough Fort, Co. Down. The arrowed ring is the year AD 1580.

# Tree-Ring Dating and Archaeology

## M.G.L.Baillie

CROOM HELM
London & Canberra

© 1982 M.G.L. Baillie
Croom Helm Ltd, 2-10 St John's Road, London SW11

British Library Cataloguing in Publication Data

Baillie, M.G.L.
    Tree-ring dating and archaeology — (Croom Helm
    studies in archaeology)
    1. Dendrochronology
    I. Title
    930.1'028'5     CC78.3
    ISBN 0-7099-0613-7

Typesetting by Elephant Productions, London SE15
Printed in Great Britain by
Redwood Burn Ltd., Trowbridge, Wiltshire.

# Contents

# Figures

# Plates

# Tables

To my father Norman Lockhart Baillie

# Preface

This book is the logical extension of a doctoral thesis presented to the Queen's University of Belfast in 1973. Although that research was carried out under the auspices of the Institute of Irish Studies at Queen's, it was and has been since carried out within the environs of the Palaeoecology Laboratory (now Centre). My thanks must go to my colleagues who have helped in many ways over the years, in particular to Dr J.R. Pilcher for his unstinting help both in the field and in the sharing of data, and to Mr G.W. Pearson, who, with his team, has been responsible for the production of all the radiocarbon results cited below. I have received invaluable assistance from Mr B. Stocks, Mr V. Buckley and latterly Mrs E. Francis and Miss E. Halliday. I owe a considerable debt to Miss J. Hillam, who processed many of the early chronologies and who latterly has supplied numerous English chronologies for comparative purposes. I wish to thank Mr B. Hartwell for his considerable help in processing the illustrations, and Dr A. Hamlin and Mr R. Warner for permission to use Plates 8 and 9 respectively. Most of the other plates were taken in collaboration with Dr J.R. Pilcher. I am grateful to Mr B. ÓRíordáin for permission to reproduce Figure 8.6 and for supplying the information used in Figure 8.5. On the production side most of the text was typed by Miss J. McKay and Miss E. Bell.

However, this work could never have been carried out were it not for the totally unselfish manner in which museum curators, archaeologists, landowners and interested parties have either supplied samples or the permission necessary to remove samples and to all of them I am most grateful. The Scottish chronology owes its existence to the enthusiasm of Mr C. Tabraham in particular. I wish to acknowledge the support I have received from the Colt Fund of the Society for Medieval Archaeology, the Royal Irish Academy and in particular the Science Research Council, without whose funding most of the more recent research could not have taken place. I hope that this volume may go some small way towards justifying their expenditure. Finally, to my wife for putting up with the most unbelievable clutter during the production stages.

# Introduction

My first introduction to the concept of dendrochronological dating was in 1966 when I was given a copy of Zeuner's book, *Dating the Past* (1958). My reaction to his section on tree-ring dating, where he described the early work of Douglass, was surprise at the elegance and simplicity of the method. In essence he was describing the basic principle of dendrochronology as shown in Figure 1. That is, overlapping of the ring patterns of successively older timbers to build a master sequence or chronology and the subsequent dating of wood samples by comparison of their individual patterns with the established chronology. The essential quality of Douglass' archaeological dating was that, in the era before radiocarbon dating, he was establishing exact ages for buildings of the prehistoric period. It is unimportant that for Douglass the prehistoric era was any time before AD 1492. In Ireland, for example, written history notionally runs back to a few centuries after Christ. However, the quality of that evidence and the paucity of references to the objects and structures with which archaeology is concerned make archaeological chronology in Ireland before AD 1100 little better than that in the south-western United States.

Curiously, in reading of the American work at that time I was not moved to ask why the method was not being applied in the British Isles. I think now that the reason must have lain in the exotic quality of the dating and the assumption that it could only work in distant desert lands. Thus I little suspected that I would later spend many years following a very similar course to that of Douglass half a century before.

The reason behind my involvement with tree-rings stemmed from a scientific background coupled with an interest in archaeology. Inevitably the possibility of working with a method which might allow the establishment of precise dates for archaeological material from periods where dating was, at best, somewhat vague (and at worst virtually non-existent) had considerable attraction. In particular in Ireland and throughout the British Isles there was a great need for some chronological backbone in the early historic period, the first millennium AD. While it may seem that it is the prehistoric period which is in greater need of dating precision, in fact this is not so. In prehistory (in the British Isles that is broadly

**Figure 1: Schematic representation of the construction of a chronology by the overlapping of successively older ring patterns.**

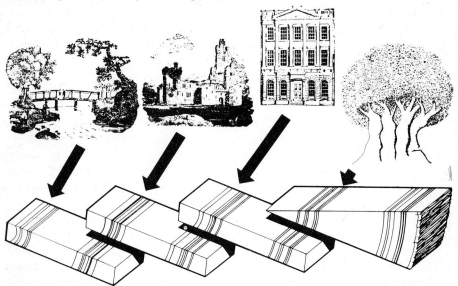

Source: Courtesy Ulster Museum.

the BC era) by definition there is nothing absolute with which to compare precise dates. It is in the protohistoric and historic periods that dendrochronology allows comparison between the dates of archaeological structures and written documentation to the greatest effect. So in the first analysis the most useful chronologies will cover the last two millennia.

But all that is jumping the gun. Before I could become involved in dendro-chronology it was necessary for someone to initiate a tree-ring project. That initiation was brought about in the north of Ireland by the availability of large numbers of bog oaks in the late 1960s. This availability of many thousands of sub-fossil oaks prompted A.G. Smith (then director of the Palaeoecology Laboratory, QUB) and J.R. Pilcher to initiate research into the possibilities of cross-dating Irish oaks. It has to be said that this was a very considerable step to take. In the 1960s botanists and palaeobotanists in England and Ireland were firmly of the belief that climate and the multiplicity of site factors affecting tree growth would make dendrochronology unworkable. In particular the claim was that because rainfall would not be a limiting factor, ring patterns would be complacent, i.e. they would vary very little from year to year. Fletcher (1978a) attributes the same general attitude to foresters and forest scientists in England in the 1960s, despite the known success of Huber's work in Germany.

So in 1968, in a generally unfavourable climate, I was employed to investigate the potential of dendrochronology using Irish oak. This volume is centred around the story of the development of oak dendrochronology at Belfast. There are, however, a number of points which are worth stressing at the outset. This volume does not attempt to cover all that has happened in dendrochronology. That subject extends to many countries and numerous tree species. The reasons behind various studies cover a host of topics from straightforward dating to geomorphological studies to dendroclimatology, art history and calibration of the radiocarbon time-scale. While it is hoped to cover an appropriate selection of useful results, in order to give the reader a feel for the subject, the text which follows is constructed around the theme of the Belfast work between 1970 and 1980. It is hoped in this way to give the reader an insight into how the method works in practice. There is almost as much to be learned from the problems and pitfalls in constructing a basic reference chronology as from studying the glossy, sometimes rather too glossy, final results of some dating exercise.

This brings me to a serious criticism of writings about dendrochronology in general. It is very easy to make the results of dendrochronological analysis seem excessively tidy. This is usually the result of attempting to present the results in too logical a fashion. The fact of the matter is that dendrochronological research is not all that logical in itself, it is only logical with hindsight. Consider the following; the closest analogy to tree-ring chronology building is a jig-saw. The pieces (assuming they exist at all, which is not certain at the outset) are scattered around as living trees, stumps, timbers in buildings and buried, either as archaeological material or as naturally preserved timbers, in bogs, river or lake beds. The pieces are accumulated not one by one but as groups of timbers, in no particular order, in the hope that some of them will be of use. The next stage is actually fitting the pieces together — constructing the chronology. Here the 'art' of dendrochronology becomes apparent. As with conventional jig-saws, some people are better at pattern recognition than others and, if the analogy is not too brutal, there are those who recognise the problems and those who might try to force the pieces together. It has to be remembered that there is only one correct pattern: each tree has grown only once and ultimately its ring pattern can only fit at one place in time. Simply because two pieces look alike does not necessarily mean that they fit together. An important part of dendrochronology is the formulation of techniques to ensure that the fit between pieces is real. So we have a jig-saw which may or may not be complete and for which the pieces are not acquired in any particular order. The result of this is that frequently the writer on dendrochronology is forced to distort the order in which things happened (in which samples were acquired and *matched*) to impose a logical sequence of events on a particular study. It is this distortion which leads to the excessively tidy nature of some reports and by extention to the mistaken belief that dendrochronology is easy. It *is* easy in concept, but it can be extremely difficult, time-consuming and frustrating in reality.

As a reaction to this too tidy, too logical picture the text which follows will go through, as far as possible, the development of the Belfast and related chronologies in the order in which events took place. Hopefully this will give a better insight into the snags and limitations of the method. It will also try to show how general pictures start to emerge at a very early stage, for example the 'gap' around AD 1350 (Chapter 11) or the clustering of the dates for horizontal mills (Chapter 9). In this vein I make no apology for my firm belief in Murphy's Law as applied to dendrochronology: 'If a tree-ring pattern can be of the least possible use — it will be.' Where this law is cited in the text it will be in this form unless otherwise stated. In short, it is to be hoped that with the aid of the following chapters it would be possible for someone, starting from scratch, to build their own independent chronology — and get it right.

While it is intended to present the chronology development pretty much as it happened, the fact has to be faced that the work at Belfast breaks naturally into sections. These sections reflect the periods for which timbers were relatively easy to accumulate, separated by periods for which there are very few available timbers. There may be a variety of reasons for this situation, but I believe that the evidence points to periods of depletion/regeneration (see Chapter 11). Be that as it may, the natural units can be listed as:

| | | |
|---|---|---|
| modern | — present to seventeenth century | — Chapter 4 |
| late medieval | — seventeenth to fourteenth century | — Chapters 5 and 6 |
| medieval | — fourteenth to ninth century | — Chapter 7 |
| early medieval | — ninth to fourth century | — Chapter 9 |
| prehistoric | — BC era | — Chapter 10 |

The reason behind laying the book out in an historical fashion is as follows. It is important for the reader to see how the reference chronologies were developed and how they were substantiated, and it is hoped that sufficient evidence is provided to allow complete faith in the chronologies described. The linking of the modern and late medieval units is gone into in considerable detail while the late medieval/medieval/early medieval links are fully justified. Because each link is independently justified the overall Irish chronologies stand as independent. They do not rely on the correctness of the German chronologies, but they do serve as an ultimate check on those chronologies.

Hopefully the reader will see how the level of cross-checking which has gone into the various chronologies is sufficient to allow them to stand as an outline for all time. They may be refined by the addition of fresh material as it becomes available, but they will not need to be moved in time. Samples dated against them will therefore be placed correctly in time. Faith in the integrity of the reference chronologies is fundamental to the nature of the results produced: at best, absolute calendrical dates fully compatible with historical chronology.

Finally, I feel there are some observations which have no other logical place in this book, for they are very much my own feelings about the subject. I record them here for what they are worth: I never cease to be amazed that denrochronology works as it does. Oak trees should not exhibit such consistently similar ring patterns; but they do. More than that it is not easy to dismiss the fact that in the British Isles we are restricted to the use of a single species for dendrochronology, on account of sample availability. Yet that species is without doubt the best subject among our indigenous trees for tree-ring analysis. So we inherit a species for which cross-dating operates and for which chronologies can be built, then we are given virtually a continuum of samples of that species back to something like 5500 BC. I say virtually a continuum because just to make things difficult a few of the pieces of the jig-saw are missing or withheld, as if someone is writing the rules as one goes along (see Chapter 9, Ballydowane/Brabstown). Of course that is only how it looks to us. Consider some Dark Age builder who cannot read or write. He fells an oak and uses its timber in some construction. The builder we will never hear from again, but that oak can for ever more tell us the exact date of its last year of growth — curious, isn't it, that oak has been given the gift of immortality?

In the text which follows, I have tried to adhere to the following convention. All tree-ring dates which are related to real time are referred to as dates AD or BC. This is comparable with historical chronology. The only other dates which warrant AD and BC are *calibrated* radiocarbon dates. Conventional radiocarbon dates and floating chronologies placed in time by them warrant ad and bc. Where there is likely to be ambiguity the terms 'real' or 'conventional radiocarbon' are used.

# Chapter 1

# Dendrochronology: an Outline History

There can be little doubt that dendrochronology as a science owes its present level of development to the efforts of A.E. Douglass in the early decades of this century. However, over the years a number of authors have busied themselves searching out examples of the use of tree-rings in the era before Douglass's classic work. Their findings indicate numerous references to the existence of rings in trees, the observation that trees record climate and the observation that trees exhibit similar ring patterns. In addition, several early authors are credited with postulating cross-dating. What is clear, however, is that it remained for Douglass to exploit the method, to establish its techniques and procedures and to build the first long chronologies which are the backbone of the science. In the sections which follow scant attention is given to the 'early' workers by comparison with that devoted to Douglass.

## Early References to Tree-rings

There is an old saying to the effect that there is nothing new under the sun. Most breakthroughs in science have built on earlier ideas, on someone else's work or previous failures. Not surprisingly, it was no different with dendrochronology and in what follows no apology is made for the blatant use of secondary sources.

It takes no great stretch of the imagination to accept that Theophrastus, a student of Aristotle, knew that fresh growth formed on the outer circumference of a tree (Studhalter, 1956, 32). Anyone seeing a stone surrounded by the folds of a living tree could have jumped to the same conclusion, whether he was Greek, Egyptian or Palaeolithic. It is certainly no surprise to read that Leonardo da Vinci had recognised the annual character of tree-rings, nor that he had deduced a relationship between ring width (for a given year) and moisture availability, to allow a weather reconstruction (Stallings, 1937, 27).

However, if Leonardo did all that, then Duhamel and Buffon, Linnaeus and Burgsdorf were positively unoriginal when in the eighteenth century they variously counted back to the ring of 1708-9 and noted severe frost injury associated with that same severe winter (Studhalter, 1956, 33). Numbers of people in the early nineteenth century appear to have been noting the similarities between ring patterns of trees cut at the same time. Studhalter champions Twining as someone who was cross-dating as early as 1827. Zeuner (1958, 400) plumps for Babbage describing cross-dating in 1837, and indeed Babbage did appear to be getting to the heart of the matter when he predicted 'the application of these views [cross-dating] to ascertaining the age of submerged forests, as to that of peat mosses, may possibly connect them ultimately with the chronology of man'. Here is someone obviously talking about the real-life nitty-gritty of dendrochronology — namely the extension of a chronology back in time by the overlapping of successively older ring patterns. If one is looking for a name to put before Douglass in a list of dendrochronologists, then that name might as well be Babbage.

## Douglass and the Early Work

Douglass did not come to tree-rings with the idea of developing a dating method for archaeologists. At the beginning of the twentieth century he was working at the Lowell Observatory at Flagstaff, Arizona. His interest lay in the relationship between solar activity and earth climate. One question, for example, concerned the various cycles observed in the sunspot numbers. Could these cycles be paralleled by cyclic phenomena in climate? Obviously if some causal relationship could be established and if the cycles in solar activity could be proven to be stable, the opportunity would have existed for the prediction of climate. The snag at the time was that while records of solar activity extended back for centuries, the records of climate were extremely short in the area in which Douglass was working. In any study of cycles it is important that the overall record be much longer than the duration of the individual cycles being observed. There would be little point in looking for proof of an eleven-year cycle (the sunspot cycle in round figures) in a ten-year series of weather observations. In order to get around this problem Douglass addressed himself to the extension of the available weather record by the use of proxy data — tree-ring records. In the American south-west large areas are what could be termed semi-arid. In such an area the growth of trees is highly dependent on the amount of available moisture, mostly rainfall. Therefore the year-to-year variations in ring width should reflect the year-to-year variations in rainfall; narrow rings being the product of drought.

Since trees grow to considerable ages it seemed possible that long records of rainfall could be reconstructed. Observation of the ring patterns of yellow pine (*Pinus ponderosa*) showed that in addition to the year-to-year variations in ring width there were trends towards wider or narrower rings over periods of years.

These longer-term trends could be clearly seen when the year-to-year detail was removed by smoothing the curves.[1] Since the year-to-year variations were due to rainfall the longer-term trends must have been due to some other factor. Since in the south-west the number of hours of sunshine per year would be essentially constant, the long-term variations must be due to variations in solar activity. With this reasoning as a background, Douglass embarked on a study of cycles in tree-rings which was to engross him for most of his life. However, cycles are not the concern of the present writer. More important by far were the developments which were to take place in the study of tree-rings as a result of Douglass's attempts to extend his proxy data records. Ultimately this led to the development of dendrochronology as a science.

Interestingly, the term dendrochronology does not make its appearance until considerably later. Douglass, in referring to the measurement of time by means of a slow-geared clock within the trees, i.e. the slow formation of rings, states that 'the term dendrochronology has been suggested as a coverall for tree-ring studies' (Douglass, 1928, 5). It is in this sense that it is currently used to describe the cross-dating of wood samples.

The first move towards the study of long proxy weather records involved the collection of samples from living or recently felled trees. It was important that the ring patterns should be firmly anchored in time and the simplest way to ensure this was to know the date of formation of the outermost growth ring. By 1909 the ring width patterns of a group of 25 pines were available. These were presented in print without being cross-identified, i.e. no check had been made to ensure that the ring patterns stayed in phase; an important factor in trees which could miss rings in years of extreme hardship. The value of cross-identification was apparently not recognised until 1911 (Douglass, 1919, 24). Thereafter cross-identification showed up various errors in the ring patterns and a tidied version was produced. By 1919 the chronology covered the period AD 1382 to 1910 (for yellow pine in the Flagstaff area).

The longest-lived yellow pines available to Douglass were limited to around 500 years. In order to investigate cycles and climatic factors over longer periods he began, around 1915, to take an interest in the giant redwoods of California (*Sequoia gigantea*). While these trees exhibited a much less specific response to climate, this was more than compensated for by their considerable life spans. However, although of great age, up to 2,000 or 3,000 years, not all of the available trees were usable and there were a number of problem rings. By 1918 a chronology for redwoods covered 2,200 years and suitable trees were being sought to extend this record. At that time the redwoods were still being exploited for timber and samples were available as wedges cut out of the stumps. Use of large samples allowed the checking of individual rings around a considerable portion of their circumference and this in turn reduced the number of problem rings. In the 2,200-year chronology only one doubtful ring remained after cross-identification of the various individual trees. This was a possible ring 'within' the ring for 1580, called 1580A. In practice this meant that in a number of trees the ring for the year 1580 included what might have been an additional ring, though in no case was this extra ring definitively present. A further sample collection trip in 1919 resolved the problem and showed conclusively that 1580A did exist. Thus 1580A was positively identified as the ring for 1580 and all of the older sections of the chronology had to be renumbered to one year earlier. By 1919 a redwood chronology of 3,221 years had been produced.

## Archaeological Extension of the Yellow-pine Chronology

Both of the above chronologies could be called long modern chronologies. Their construction had required little in the way of extension by cross-dating since they were composed almost exclusively of recently felled trees. By 1923 renewed interest in the Flagstaff yellow-pine chronology yielded an extension back from 1382 to 1284. This appeared to exhaust the possibilities of the living trees. To go back further in time it became necessary to study timbers from prehistoric Hopi villages and ancient ruins in the area. The semi-arid conditions were responsible for the long-term preservation of wood on these sites and the timbers could be sampled either by slicing (where appropriate) or coring.

By 1922 a floating chronology had been constructed using timbers from Aztec, New Mexico and Pueblo Bonito some 50 miles to the south. This chronology covered RD (Relative Date) 230 to RD 543, a total length of 314 years. There was no way of knowing the true age of this chronology except by joining it to the 1284 modern chronology. Since no join could be found, Douglass had to assume that RD 543 was some years older than 1284. The importance of consolidating this chronology was obvious. Once tied down in real time, the dates of these prehistoric sites could be established very accurately. In addition, the chronology available for cycle studies would be extended. The particular attraction of being able to date the archaelogical ruins seemed worth a considerable effort. In the mid-1920s Douglass (1928, 61) was able to hypothesise two ways to achieve the dating of the Aztec-Bonito complex. The first would be to bridge the gap by overlapping successively older timbers until such time as a link with the floating chronology was achieved (a procedure which is considered classic dendrochronology today). The second was by what he called 'sequoia comparison'. Since the sequoia chronology ran continuously back to around 1200 BC it had to cover the period of the Aztec-Bonito floating chronology. The question was, could a yellow-pine chronology from New Mexico be cross-dated against a sequoia chronology from California? Douglass reckoned that the best hope had to lie with the former, the bridge method. This was on implicit recognition by him of the questionable character of tele-connection, i.e. the matching of ring patterns over very long distances (of which more later).

## Cross-dating

At this stage the discussion had moved away from the study of essentially living, or at least recent trees, to the study of undated individual ring patterns. The only way to place such ring patterns in time was by cross-matching them with others. It is important to understand the concept of cross-dating as developed by Douglass. A good example of the tying down of an undated sample took place as early as 1904. Studying the ring records of a number of pines newly felled near Flagstaff in that year, Douglass noted that a group of narrow rings occurred in each tree around 1880. In particular the rings for 1880 and 1883 were very narrow. To quote Douglass (1937, 3):

> While this was fresh in mind, curiosity raised the question whether this same [narrow] group could be found near the outside of a slightly weathered stump whose date of cutting was unknown. On examining the rings in the stump the group was found at once but was only eleven rings in from the outside. So that tree must have been cut in 1894. The owner of the land was sought and asked when his timber was cut and he answered, 'in 1894'.

This is almost certainly one of the first examples of the dating of an unknown specimen by dendrochronology.

Here then was the key to the methodology behind cross-dating in the south-west. In any piece of wood to be dated, configurations of rings were sought which were known from previous work to occur over a unique set of years. For example, a set of four narrow rings followed by two wide rings with a noticeable narrow seven years later occurred over the years 1215 to 1227. A wood sample exhibiting this configuration fifteen years in from its bark surface could be assumed to have last grown in 1242. Final confirmation would be obtained by checking the remainder of its ring pattern for further extreme rings coincident with the known chronology. This same approach could be applied to substantial pieces of charred wood or charcoal from archaeological excavations, provided only that they contained sufficient numbers of rings to allow definitive matching. Douglass is attested to have retained a prodigious number of such 'signature' patterns in his memory with the result that he could often date archaeological samples simply by looking at their ring patterns under a lens and comparing these with his memorised chronologies.[2]

Obviously the whole programme of research was not simply carried about in Douglass's head. In order to facilitate the recording of the patterns of rings the following scheme was devised. A horizontal scale in years was employed. Plotting from left to right, from the centre to the outside of the sample, each narrow ring

**Figure 1.1: (a) Skeleton plot and graphical representation of the ring pattern for the period AD 1215 to 1242 in the American south-west. (b) How HH-39 formed the key link between Douglass's modern and archaeological chronologies in 1929.**

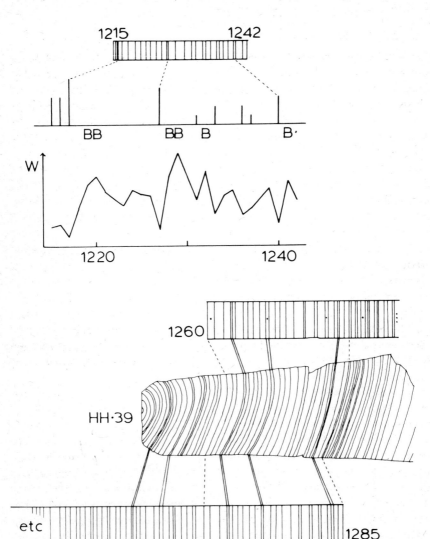

was assigned a vertical line; the narrower the ring the longer the line. Very wide rings were denoted by a capital B under the horizontal line. Average rings were not marked. Figure 1.1(a) shows how the configuration for the period 1215 to

1242 appeared when plotted in this format. Because only the extreme rings were denoted, this representation became known as the skeleton plot. In practice a master scheme was created against which the skeleton plot of each new sample could be compared.

## The First Decisive Archaeological Cross-dating

As noted above, Douglass had made the decision to attempt the absolute dating of his Aztec/Pueblo Bonito 314-year chronology by bridging to the relatively local Flagstaff yellow-pine chronology which extended back to 1284. His decision not to try tying down the floating chronology against the very long sequoia chronology was probably conditioned by both the distance between the sources and the differing climate and species responses. (If he could complete the pine chronology independently, then it could be compared at one specific position in time against the sequoia chronology. Any cross-matching found at that unique position would, if it occurred, confirm the correctness of his archaeological chronology. On the other hand, since cross-dating could not be assumed, any attempt to date the archaeological chronology directly against the sequoia chronology might well have led to a spurious result.)

The situation in which Douglass found himself was a classic one in dendrochronological terms. He had two yellow-pine chronologies whose exact relative placement in time was unknown. The floating chronology could be surmised to be older than the long modern chronology on the basis that no cross-matching could be found between them. It must be remembered that any overlap between the chronologies would have to be long enough to allow definitive cross-dating, a short overlap being no better than no overlap at all. The options in this situation were twofold. Attempts could be made to extend the modern chronology back in time to tie up with the floating chronology or vice versa.

In order to achieve one or other of these aims an expedition was sponsored by the National Geographic Society in 1923 to collect material for chronology extension. This was named the First Beam Expedition. Large numbers of beams were collected from structures of all ages. These included pueblos, historic buildings and prehistoric sites within the south-west. It was hoped that long-lived timbers from some of the early historic sources might match with and extend the modern chronology back to before 1284. Unfortunately, although many timbers could be dated against the existing chronology, none extended it significantly back in time (an analogous situation is discussed in Chapter 7). On the other hand, many early timbers were found to cross-date with the floating Aztec/Pueblo Bonito chronology. In particular, timbers from a ruin in Wupatki, north-west of Flagstaff, were tied to the floating chronology. Other timbers from this same ruin formed the basis of a second floating chronology. Importantly, on the basis of accumulated pottery evidence from the various cross-dated sites, this second chronology of approximately 140 years could be placed later than that from Aztec/Pueblo Bonito, i.e. between it and the modern chronology. To quote from a recent review by Robinson (1976, 11):

> As analyses proceeded on the collections made by the First Beam Expedition, more and more specimens gradually yielded to cross-dating within one or other of the two relative chronologies. These two were ultimately merged, in 1928, to form a single chronology of prehistoric ruins with a length of 585 years. The status, then, of chronology building early in 1928 consisted of two long records. The absolute chronology extended from the present back to about 1400 with confidence, and weakly — because it was based on few trees — to about 1300 [in fact to 1284]. The other was a floating chronology of 585 years of unknown absolute age based on specimens from approximately 30 prehistoric ruins.

Clearly the First Beam Expedition had not achieved the desired result, consolidation of the floating chronology. However, it had been responsible for considerable progress, in particular extending the floating chronology forward in time. This prompted the organisation of the Second Beam Expedition in 1928. Since it was known that some material from Hopi villages had cross-dated at the older end of the modern chronology, i.e. in the fourteenth and fifteenth centuries, a major collection of early Hopi timbers was implemented. As discovered elsewhere with random sampling in gap-bridging attempts, the law of diminishing returns sets in (see Chapter 11). Of the many samples cross-dated in the 1928 collection most matched with the existing chronology; only one gave an extension back to 1260.

Random sampling was of course a useful technique for getting an overall picture. On the other hand, in order to find timbers of specific age it was necessary to adopt a more systematic approach. Fortunately, this was possible as one outcome of all the dating activity had been a better understanding of the archaeology of the structures being dated. (The relative dates of at least thirty sites were known even if their calendar dates were not.) This was important because it enabled the archaeologists to suggest where timbers of particular ages might be found. The key to the selection of suitable sites for study lay in the understanding of the pottery sequence of the overall period.[3]

Putting together the relative dating information derived from the tree-ring cross-datings and the available pottery evidence, it became clear to the archaeological collaborators that a general picture was beginning to emerge. Sites dating to the later part of the 585-year floating chronology were characterised by red background polychrome pottery. On the other hand, sites dating against the oldest part of the living chronology had orange and yellow background pottery

(Robinson, 1976, 12). The archaeological suggestion was that the sites most likely to provide bridging timbers would be those where orange pottery was predominant (Haury, 1962, 12). However, the sites producing pottery of the correct colour were as a rule not the best sites for timbers. Most were timberless or occurred in areas where it was unlikely that the right sorts of timber would have been used by the original builders. This effectively reduced the number of 'orange' sites likely to be of use. Despite this, excavations were undertaken at a number of sites in 1929. Sampling had to be through the medium of excavation for the simple reason that in these timberless ruins only charred beams were likely to have survived.

One of the sites chosen for study was the Whipple ruin at Showlow. It fulfilled all of the criteria for a possible bridge site. On the fifth day of excavation, 22 June 1929, a charred roof timber turned up with most of one radius preserved. Douglass arrived at the site in time to assist with the lifting of the beam. Preliminary examination of the ring pattern of this timber (labelled HH-39 on the numbering system devised by the excavators Hargrave and Haury) showed that its outer rings ran out to around 1380 and extended back to 1237 (Douglass would not have been the only happy person on the site that day as a $5 bonus scheme was in operation for anyone finding a sample with more than 100 rings). The question was, would this timber cross-match with the floating chronology? Later the same day Douglass was able to pronounce that it did. There was a recognisable overlap of 49 years between HH-39 and the floating chronology, specifying the 585th year as 1285. Figure 1.1(b) shows schematically what this match looked like. In fact an overlap between 1260 and 1285 had already existed, but it had been too short to allow definitive matching (Haury, 1962, 12-13).

As soon as the cross-dating was verified all the relative dates, associated with what had been the floating chronology, were converted to absolute dates. At a single stroke Douglass had given the archaeology of the south-western United States a precise calendrical framework for the first millennium AD. Absolute dating precision was at once routinely available in advance of anything available elsewhere.

This archaeological chronology back to 701 was to be refined and extended in essentially the same way as outlined above. A description of the steps involved in the extension of the chronology back to the BC era is given in Robinson (1976, 15-16). At the present time the furthest extension back in time of the tree-ring chronology for the south-west is 322 BC.

The true importance of this work lay in Douglass's confidence in building a chronology back into prehistory which relied on nothing but the cross-matching of tree-ring patterns. The datings he produced were not conditioned by archaeological considerations but were truly independent.

## The World's Longest Chronology

The 3,000-year chronology developed for the giant redwoods had been put together in a comparatively short time. This was a reflection of the enormous life spans of the sequoias. The construction of the chronology required only the ironing out of inconsistencies in the individual ring patterns, using replication. The construction of the yellow-pine chronology had taken much longer due to the necessity for considerable numbers of overlaps, the individual pieces having to be found and matched together.

Before leaving the Americas it is necessary to mention the work which has taken place on the bristlecone pine (*Pinus aristata*). During the decades after 1929 considerable effort was put into both archaeological dating and climatic reconstruction. Trees were sought which would add to the understanding of the past climate of the United States. This may have been prompted in part by the vicissitudes of the 1920s and 1930s, but was probably conditioned by the known variability of climate in the recent past. Trees which grew close to the extremes of their habitat, for example close to the upper or lower tree lines, should be particularly sensitive recorders of climatic variation.

In searching for suitable long-lived species for such studies, it was inevitable that the bristlecones would eventually be discovered. Ferguson (1968) credits Schulman with first drawing attention to these trees in 1953. The bristlecones grow in a number of areas in the western United States, in particular in the White Mountains of east-central California. There are two very significant factors. First, these trees grow at very high altitudes, 3,000 to 3,350 metres. Second, they reach incredible ages. Examples of living specimens up to 4,600 years old have been recorded (an example in Nevada has been credited with a span of 4,900 years). In addition, the conditions in the high altitudes where these trees grow, combined with their resinous nature, favour preservation of the wood for long periods after tree-death. Stumps and weathered remnants (which died up to thousands of years ago) can be collected from the ground surface. Thus the raw material was available for the construction of a chronology covering many thousands of years.

The construction of a definitive bristlecone chronology was, however, not without its difficulties. The trees grow extremely slowly, and examples showing 40 rings per centimetre are common. With such narrow-ringed material, years of particular stress result in rings being locally absent. In fact in any one core as many as 5 per cent of the total number of rings may be missing. In order to overcome this problem there was a strong need for multiple cores and for replication between trees. The requirement for absolute accuracy within the chronology meant that, although the individual trees exhibited ring patterns up to four millennia in length, by the time of Schulman's death in 1958 the chronology had been checked only as far as 780 BC. In the following decade the work was taken over by Ferguson and by 1969 a chronology of over 7,000 years had been completed (Ferguson, 1969, 3). Subsequently this was extended to 8,200 years. While at least the possibility existed of checking the first 3,000 years of this chronology using the sequoia chronology constructed by Douglass, it would be reasonable to ask how it is known that the remainder of the chronology is correct? Obviously replication is the prime method of ensuring accuracy. However, an independent check on the last 5,405 years of the chronology was possible using a separately developed bristlecone chronology for the Campito Mountain region of California (LaMarche and Harlan, 1973). This work showed that the two chronologies agreed exactly over the whole of the last five millennia. It is therefore safe to assume that the whole of the chronology is correct.

The primary importance of this chronology undoubtedly lay in its contribution to the calibration of the radiocarbon time-scale (see Chapter 12). It is possible that this calibration may be extended still further back in time given the large quantities of sub-fossil remnants of these trees as yet unexploited. In particular the known occurrence of samples 9,000 years old (dated by radiocarbon only) lends hope that in time an absolute chronology may be available covering at least 10,000 years (Bannister and Robinson, 1975, 220).


## Europe

As indicated above, various European workers had studied growth rings in trees, in particular in the eighteenth and nineteenth centuries. Figure 1.2 shows one eighteenth-century author's idea of oak growth (the plate comes from an unknown source). Clearly the order of growth was understood even if the timber in question was somewhat complacent in ring width. However, resolving the question of the number of rings per year hardly classed as dendrochronology. The most inspired piece of early European work appears to have been by Kapteyn in the 1880s (Kapteyn, 1914). He measured multiple radii from a selection of groups of oaks in Holland and Germany and then used cross-dating to confirm the correctness of the ring patterns. Having thus replicated his data, he proceeded to construct mean site chronologies which were then compared from area to area. In addition, he investigated the correlation relationship between the mean curves and existing weather records. In all, his chronologies extended back for several centuries and his work must place him pretty firmly in the forefront of European dendrochronology.

In 1912 and 1913 Douglass collected pines from England, Scandinavia and Germany and processed them, principally as part of his cycle research. Indeed most of the early-twentieth-century work on European tree-rings was related to

Figure 1.2: Eighteenth-century illustration of oak growth.

Source: The unprovenanced plate kindly supplied by Mr R. Howarth.

studies by foresters, biologists and meteorologists. It was not until the 1930s that investigations into dendrochronology as a dating method were begun by Huber in Germany. Huber was convinced that it would be possible to apply Douglass's methods to the resolution of problems in medieval dating in Europe. Inevitably oak was to form the mainstay of chronology building since it represented the prime building material in most of north-west Europe. While other species were successfully exploited for dating purposes, including fir and beech, there can be little doubt that the construction of the German oak chronology formed the natural extension to Douglass's work.

Huber was able to use long-lived Spessart oaks to push an initial 'living-tree' chronology back to the fifteenth and, later, to the fourteenth century. From there it was inevitable that bridging would have to be used for further extension. Here lies the fundamental difference between the problem faced by Huber and that which had earlier confronted Douglass. Whereas Douglass had been pushing back into the unknown, Huber was able to draw on a very large number of well dated sources of medieval oak. In Europe it was possible to go directly to sources of sixteenth-, fifteenth- . . . eleventh-century timbers and by so doing Huber and his associates were able to complete a 1,000 year chronology by 1963. Using this initial chronology, it was then possible to embark on a programme of dating buildings, archaeological sites and art-historical objects. In the process the essential methodology for dealing with oak was worked out. For a full account of Huber's work see Liese (1978).

In Europe, using temperate tree species, the skeleton plot was found to be unsuitable, as it ignored too much of the information available within the ring patterns. Signatures undoubtedly did occur and were recognised by Huber. He suggested that any year in which greater than 75 per cent of all trees showed a similar trend towards a narrower (or wider) ring could be termed a signature. Such years were marked, not as in the skeleton plot, but simply as heavy lines on an overall mean ring pattern. In addition to individual signature years it was noted that certain signature patterns could occur consistently. The most famous of these was the so-called 'German Saw', a regular two-year cycle of minima and maxima from 1530 to 1540 (Huber and Giertz, 1970, 220). Huber recognised that the cross-dating of individual ring patterns had to be carried out using the correlation between the whole of the ring patterns. It was not sufficient to rely on the signature years or patterns alone. This led to the development of the 'coefficient of parallel variation' as a statistical back-up to the visual matches, something which Douglass had not found necessary, given the remarkable consistency of the patterns in the American south-west.

Huber had successfully adapted the ideas of Douglass to the European environment. He had worked out the methodology for oak dendrochronology in central Germany and had successfully applied the method to the solution of problems in medieval dating. In terms of straight chronology building he was overtaken by Hollstein working at Trier, principally on oak material from west of the River Rhine. By 1965 Hollstein had published a chronology back to AD 822, but, more importantly, he had roughed out the elements of a continuous chronology back into the first millennium BC using diverse archaeological sources from Iron Age houses to Celtic and Roman bridges and Dark Age coffins. It was to take Hollstein a further fourteen years to complete this chronology back to 700 BC, and by that time he was only one of a number of German successors to the throne of Huber (Hollstein, 1979).

Huber, of course, had done much more than simply construct a 1,000-year chronology. He prompted the investigation of the relationship between tree growth and climate as well as analysis of the variations in ring width with geographical location (though of course in this respect he was only following in the footsteps of Kapteyn). In addition, he was a dedicated researcher into the physiological responses of trees and overall his place in the history of European dendrochronology is assured. However, as a scientific investigator it is certain that he would have acknowledged the primacy of Douglass and indeed might have been less than enthusiastic about the promotion of a Huber 'cult' as exemplified by the proceedings of the 1977 Greenwich Conference.


**Novgorod, a Differing Approach**

It is not intended to cover the early history of dendrochronological dating area by area around the globe. It is, however, worth noting one major dating exercise which was undertaken in Russia around 1960. Whereas in America Douglass had used totally independent bridging to tie down his floating chronology and in contrast Huber had been able to draw on well dated souces of timber to construct his 1,000-year sequence, this Russian work used a different approach to establish the age of an archaeologically derived chronology.

The work was carried out on timbers from the medieval city of Novgorod, from levels belonging to the tenth to fifteenth centuries AD. Importantly, and in contrast to the rather individual character of most tree-ring projects, the Novgorod work was carried out in a tree-ring laboratory set up specifically by the Institute of Archaeology of the USSR Academy of Sciences. This in itself shows an implicit realisation of the importance of the method all too rarely recognised elsewhere. By cross-matching the ring patterns of pines from successive archaeological levels it was possible for the dendrochronologists to build a continuous 579-year floating Novgorod chronology. On archaeological grounds the chronology had to cover the approximate period 900 to 1450. It was known that little likelihood existed of finding long-lived modern trees to bridge back to the floating chronology. In addition, it would have been time-consuming to construct such a bridging chronology by conventional means. So another route was chosen which relied on the large amounts of historical documentation associated in particular with the building dates of churches.

In Novgorod wood samples were obtained from five well documented standing churches. The timbers from the five were dated against the floating chronology and the *relative* dates of the outermost rings obtained. In each case it was assumed that the last growth ring of the sample had been formed in the year before the foundation date recorded in the chronicles. The building dates were 1300, 1355, 1384, 1418 and 1421. It was found that the intervals between the historical building dates were the same as the intervals indicated by the relative tree-ring matches (see Figure 1.3). It was thus possible to assign the chronology to the period 884-1462 (Kolchin, 1962). In order to date a floating chronology in this way it was necessary to have extremely reliable historical information. It is doubtful if such refined historical dating would be available for any set of timber structures in the British Isles and a dating exercise of the Novgorod type would be hazardous.[4]

One dating exercise associated with the Novgorod excavations which is particularly famous concerns the rebuilding (or relaying) phases of the wooden pavements (corduroy pathways) of Velikaya Street. A classic archaeological sequence of 28 successive pavements was excavated. Each represented a sealed layer, an archaeologist's dream, and archaeologically the questions were (a) what length of time had elapsed between each rebuilding and (b) what were the dates of the sealed layers? It was possible in practice for the dendrochronologist to assign calendar dates to all 28 of these layers from the earliest, laid down in 953, through all the successive intervals which varied between 12 and 20 years, up to 1462 — surely the ultimate control on the chronology of a site. Interestingly, in at least ten cases, the laying of new street levels coincided with, or closely followed, the dates of fires recorded in the chronicles and this was supplemented by a further six refurbishings associated with archaeological evidence for previously unrecorded fires (Thompson, 1967, 33). This represents an excellent example of the full compatibility between tree-ring dates and history leading to a better understanding of an important site.

## Ireland

Overall there seems to be little doubt that dendrochronology can offer excellent dating control provided that it is treated logically. The following chapters tell the story of the construction of an independent chronology in a new area on the Atlantic seaboard. First, there is a little detail regarding the raw material available

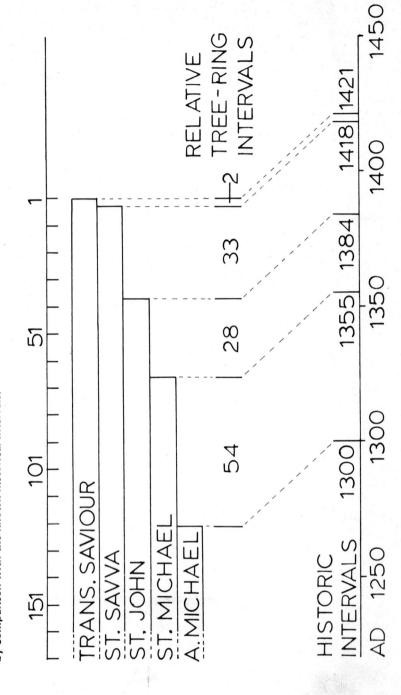

Figure 1.3: How the relative placement of the tree-ring patterns from Novgorod allowed specification of the chronologies by comparison with the known historical intervals.

for such a study in this new area and on the practicalities of working with it. This
is not a textbook on wood physiology, tree growth or statistics, but the scene is
set with some basic information, enough hopefully to satisfy the non-specialist.
What follows should adequately demonstrate the way in which the essential
independence of the method can be ensured. If this can be achieved on an island
as small as Ireland it will bode well for the construction of such chronologies in
less remote areas.


## Notes

1. Smoothing can be achieved in a number of ways. A simple example would be to re-
plot the ring pattern by substituting for each ring width the mean of the five ring widths of
which it is the central value. This would be a five-year running mean.

2. Douglass was working before the advent of the electronic computer. It is interesting
to note that subsequent checking of his dates, using computer correlation routines, failed to
find an example where he had been in error (Bannister, personal communication).

3. People who have not visited prehistoric pueblo sites will have little comprehension of
the quantities of pottery sherds which can occur in and around these structures. The dry
conditions favour preservation so that coloured and decorated fragments abound.

4. In an unfortunate series of attempts at something similar for early medieval sections
of English chronology, Schove failed to date timbers from Old Windsor correctly. This was
at least partially due to the use of inferior evidence derived from coins, weather records and
radiocarbon dates (Schove, 1979). In the British Isles ring patterns should only be placed in
time by definitive cross-matches with other dated patterns.

# Chapter 2

# Oak Growth and Ring Measurement

In this chapter it is intended to look briefly at the reason behind the choice of oak as the principal species for tree-ring analysis in Europe. Oak has a number of advantages for dendrochronology, including a clear annual character, long life and easily distinguished outer wood. This latter factor can be critical for accurate dating, as will be seen below. Oak is not without its problems, however, and these are looked at in some detail.

## Choice of Oak

In the British Isles the dendrochronologist, in attempting to build long chronologies for historic and archaeological dating, is constrained to concentrate on oak, *Quercus robur* and *Quercus petraea.* Both species are indigenous to Britain and Ireland, but in general, with the exception of living trees, it is difficult to distinguish between the two. In living trees observation of the leaves and acorn cups may allow separation, though this can be complicated by the existence of hybrids. Walker (1978) discusses separation on the basis of wood anatomy but this holds up only for relatively wide-ringed material. In the Irish work, and indeed in most work within the British Isles, no serious attempt has been made to identify individual timbers to species. This is particularly true of ancient timbers, where such identification is often impossible. The failure to identify to species has had little noticeable effect on the building of chronologies which are simply generalised to oak. The concentration on oak as the principal species for dendrochronological study in the British Isles is purely the result of sample availability. In Germany and Switzerland, where other species such as pine or beech are available as timbers in buildings, it has been possible to build chronologies back for hundreds of years. In the British Isles chronologies for pine, beech, elm and ash could be constructed back to the seventeenth century (using long-lived modern examples), but the rarity of these timbers in buildings would make it impossible to construct longer chronologies. Oak, on the other hand, has always been used as a building material, back to prehistoric times, and samples

45

are available for almost all periods in the last two millennia. (In Ireland the occurrence of large quantities of sub-fossil or 'bog' oaks extends the potential availability to around eight millennia (Pilcher, 1973).

It should be noted that the paucity of species other than oak does not preclude the analysis of floating groups of timbers such as the medieval elms from London (Brett, 1978, 195). Similarly, it is known that large quantities of ash timbers are available from the Viking levels in Dublin. In both cases it would be impossible to build independent species chronologies back to join with these floating groups. It is, however, conceivable that they could be dated against existing oak chronologies, especially if samples of both species occurred in contemporaneous structures.

## Oak Growth[1]

Amongst European timbers oak is readily distinguished from all others by being ring-porous and having two sizes of rays. The larger rays are easily visible to the naked eye, as are the individual vessels of the spring-wood. These two criteria simplify the identification of oak in the field (Jones, 1959). In common with all trees growing in a temperate climate oak develops one growth increment or ring per year. Thus under normal circumstances it leaves, over its lifetime, a record of the number of years it has been growing. This record is preserved as a series of concentric rings exhibited in a cross-section of its main stem. The annual rings develop immediately under the bark and are due to the growth and division of a thin layer of cells called the cambium. In oak and in a limited number of other dicotyledons (elm, ash, etc.) there is a sudden change in character between the spring-wood and summer-wood of a total season's growth. The spring-wood consists of large vessels formed during the period of shoot growth which takes place between March and May (see Figure 2.1). Since this is before the establishment of any great photosynthetic leaf area, most of the energy and raw materials for the formation of spring-wood must come from food stores laid down in the previous year (Varley and Gradwell, 1962).

The spring-wood vessels are thin-walled and are responsible for a purely physical upward translocation of water from the roots to the extremities of the aerial parts. At the time of leaf expansion in oak, i.e. some time in mid-May, hormonal activity dictates a change in the quality of the xylem and the summer-wood becomes increasingly fibrous and contains much smaller vessels. Plate 1(a) shows a transverse section of oak wood. The horizontal bands of large vessels represent the spring-wood and the dark bands the summer-wood. During the production of the summer-wood, the cambium oscillates between two types of xylem production, i.e. dark bands of fibres interspersed with pale bands of small summer-wood vessels. These can be clearly seen in the uppermost ring in Plate 1(a).

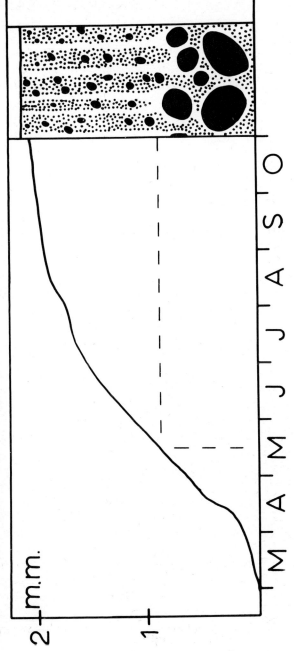

Figure 2.1: Internal structure of a growth ring in oak and its development.

Source: After Varley and Gradwell (1962).

Plate 1: (a) Typical oak cross-section showing the narrow rings associated with the years 1816 and 1817 (in Ireland).

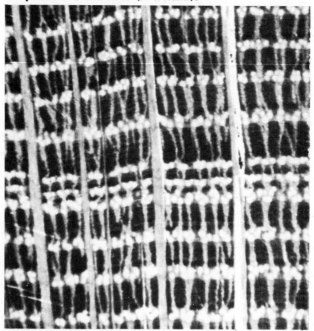

Plate 1: (b) Wide spring vessels within the ring for AD 1543 in QUB 361 giving the impression of an additional ring. This could be termed a 'ghost' ring.

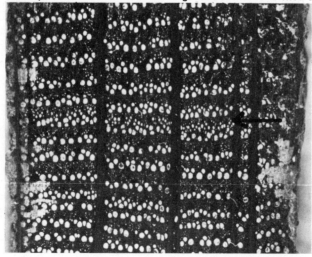

Plate 1: (c) Local absence of spring vessels in an Early Christian sample from the Little Island horizontal mill, Co. Cork.

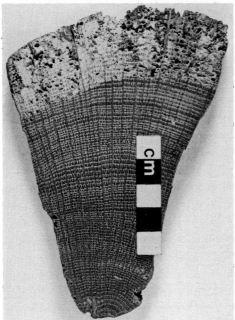

Plate 1: (d) Contrast between the heavily attacked sapwood and the resistant heartwood in a typical medieval building timber. The outer curved surface represents the 'waney' edge and can be identified as the felling year.

Plate 1: (e) Local presence of heartwood extending to the felling year. In such a case the addition of a sapwood allowance would produce a felling date much later than the true date.

Plate 1: (f) The effect of sudden defoliation (?) on the ring pattern of a tree. Immediately after the event the spring-wood width is reduced to a single line of vessels (see also Figure 3.1(c)).

This variation between the spring- and summer-wood in oak has associated with it a variation in density. The spring-wood has generally larger cells and proportionately a smaller amount of wall substance per unit volume. This variation in density and hence in colour makes the transition between the end of the summer-wood of one year and the beginning of the spring-wood of the next year very distinct. The conclusion of a season's growth is marked by the formation of a line of small and dense cells at leaf fall. The tree is then dormant over the winter period until growth is triggered in the spring of the following year, presumably by temperature-induced hormonal activity.

Esau (1960) states that the width of rings is easily influenced by the external environment and is therefore variable. In oak, as stated above, the width of an annual growth increment is made up of two distinct portions: the spring-wood, which is produced during leaf expansion and is dependent largely on the food reserves of the tree, and the summer-wood, which is dependent on the immediate food supplies of a particular year.

The large vessels of the spring-wood of oak are comparatively uniform in size and occur in bands usually only one or two vessels wide. Because of this, the width of the spring-wood in oak varies very little from year to year in comparison with the variations in the width of the summer-wood. The reasons for variation in the summer-wood width are complex, but clearly must be dependent on the photosynthetic food supply available to the tree. This in turn is dependent on the foliage canopy, the temperature and the amount of sunlight available for conversion into food. The fact that spring-wood is almost invariably present for every year of the tree's growth, while summer-wood can be totally absent, indicates that the former is a much more certain process. In the case of oak, where the supply of water to the leaves is not dependent on the vessel-width in a particular year, the width of the summer-wood is to a large extent independent of the width of the spring-wood.

From this we can deduce that the stored information available to us in a study of the tree-rings of oak is contained almost entirely in the summer-wood. In the case of two consecutive tree-rings, if we balance the infinitesimal narrowing of the second ring due to the increasing diameter of the trunk with the increased capacity for growth resulting from the yearly foliage increase, then the quality of the growing season of the second ring compared with that of its predecessor should be shown by its relative width. That is, if the second season had poorer conditions for growth, the second ring will be narrower and vice versa. One significant exception of this is the case where defoliation of the tree in early summer leaves no record of summer growth. In this case the defoliating agent effectively obscures the overall growing condition of the year.[2]

## Missing Rings and Double Rings

Lowson (1966, 100) states that the production of one growth-ring each year is not a certain process. It is possible, under certain conditions, for a tree to miss out a growth-ring or to produce two growth-rings in one season. The first of these phenomena occurs where there is insufficient xylem growth to form a noticeable ring, normally due to extreme drought. The second is the result of a defoliating agent which arrests the tree's growth early in the year after which there is a secondary out-growth of new foliage (Esau, 1960, 246). The cause of the defoliation could be drought, late frost or insect attack. These remarks apply generally, although Lowson also states that 'under British conditions, age estimates based on ring-counts are subject to only a small amount of uncertainty'. This last statement can be confirmed by the dendrochronological work on oak throughout Europe and the British Isles. In fact it is extremely rare to find an oak which has missed a year's growth or which shows two distinct rings in a single year. However, cases do arise where narrow bands of rings or wide spring-wood cause measurement difficulties. Both of these cases are dealt with later in this chapter. These measurement difficulties generally arise in attempting to interpret the ring record of a single tree. They are eliminated in all cases by study of the ring records of other trees growing over the same period, i.e. by replication.

Reasons for the reliable recording of annual increments may be found in the ring-porous nature of oak. A ring can be visually resolved in oak provided the line of large spring vessels is distinct from those of the adjacent years. Since hormonal activity dictates a change in the xylem production around the middle of May, when the leaves are fully opened, a distinct line of large vessels will be in existence by that time. We know that up to mid-May the tree does not have a large photosynthetic leaf area and must be growing on the strength of its reserve food store. Thus production of the spring-wood is governed only by the need for a photoperiodic/temperature reaction to trigger cambial activity. Since this triggering will take place each year with total certainty, the production of spring-wood each year is equally certain. The relationship between spring growth and the food reserves of the previous year is shown clearly in the reduced width of the spring-wood in the year following a catastrophic defoliation.

It is more difficult to establish a theory for the lack of double rings. However, it seems likely that an oak uses a great deal of its reserve food store in establishing its foliage canopy. In the event of partial defoliation due to frost or insect attack the tree struggles on through the growing season building up what reserves it can for the following year. There are no records of whether defoliation is total or partial and in general it is impossible to relate a sudden reduction in tree growth to any specific agency. Overall, providing that the bands of spring-wood of successive rings can be satisfactorily resolved, the ring record of oak forms an accurate annual time-scale.

## Anomalies

It would be misleading to suggest that the ring patterns of oak trees are immutable calendars. Occasional anomalies do occur. Huber and Giertz (1970, 209) illustrate what can best be described as a 'ghost' ring in a German oak. This is something which looks like an annual ring but which is in fact only internal structure within a genuine ring. Plate 1(b) shows a similar phenomenon in a single Irish sample, QUB 361. This configuration, which looks to the naked eye like an actual ring, is simply a series of abnormally large summer vessels within the ring for 1543. As far as is known there is no record of a genuine double ring configuration in oak.

Similarly, completely missing rings are unknown in oak, i.e. a ring which is absent around the whole circumference of a tree. Such an event, if it did occur, could only be identified by implication. Such a situation would be where, in a definitive matching position, a portion of a ring pattern clearly went out of phase with no apparent cause. However, there are recorded cases of locally absent rings. These can be anything from a few missing vessels to a situation such as that described by Morgan in an informal Tree-Ring Society newsletter in March 1979.[3] She observed in a Bronze Age oak stake from the Somerset Levels (*c*.1000 BC) a ring whose early wood (spring vessels) was absent over some 5 cm of circumference. The ring over the remainder of the circumference appeared to be perfectly normal. Plate 1(c) shows a minor occurrence of such a local absence in the ring for AD 458 in QUB 3671, a sample from Cork.

## Sapwood in Oak

We have seen that oak has many advantages for dendrochronological analysis, clear ring record, long life and reliable annual character. In addition, it has the important advantage that the sapwood is clearly distinguishable from the heart-wood. In any dendrochronological study aimed at the establishment of accurate calendar or relative dates, it is a requirement that the felling dates of the trees in question should be accurately estimated. This requirement can only be fulfilled successfully where the outer growth ring of a particular tree is present. Ancient timbers will often have suffered loss of rings due either to physical deterioration or the removal of wood in a shaping process. Fortunately, with oak it is often possible, even with some of the outer wood missing, to obtain an estimate of the felling date.

In woody plants there are three basic elements of the secondary xylem. These are the tracheary elements or vessels, concerned with the movement of water, the fibres concerned with support and the parenchyma cells concerned with the translocation of food. The first two of these, the tracheary elements and fibres, are made up of dead cells, and only the parenchyma cells are alive. The portion of the tree containing live parenchyma cells is called the sapwood as distinct from the totally inert heartwood. The dying of the parenchyma cells is preceded by numerous changes in the wood which visibly differentiate the active sapwood from the inactive heartwood.

The differences between heartwood and sapwood are not primarily concerned with strength, the fibres being dead in both cases, but are basically chemical. With increasing age the wood loses water and stored food and becomes infiltrated with organic substances such as oils, gums, resins and colouring agents. In oak, tyloses develop in the tracheary cells, blocking the vessels. These various changes render the heartwood more durable, less penetrable, less open to attack by decay organisms and visibly distinct from the living sapwood. Further, the loss of food supply makes the heartwood unattractive to wood-boring insects. Plate 1(d) shows an example of complete sapwood which has suffered severe insect attack.

The proportion of sapwood and heartwood and their differentiation are highly variable in different species and in different conditions of growth. In oak, however, the sapwood is distinctly different from the heartwood on two counts. First, it is much lighter in colour and, second, the large spring-wood vessels of the sapwood are hollow, while those of the heartwood are blocked by tyloses. Further, and most important from the dendrochronological viewpoint, the number of growth rings which make up the sapwood of an oak is to some extent predictable.

*Sapwood Estimation*

The oak timbers which occur in historic and archaeological contexts have generally suffered at the hands of a woodworker. It is unusual to find a complete oak trunk used in buildings or structures of any period. Apart from the loss of rings due to woodworking processes, the removal of oak bark for tanning (McCracken, 1971, 79) must often have resulted in the adze removing some of the outer sapwood years. While in theory the only way to tell the outside ring of a tree is by the presence of bark, in practice there are two cases where the certain presence of the outer ring can be established.

(1) Where a beam retains the original curved surface of its sapwood, observation of a transverse section should show the outer ring continuous over a significant arc of the circumference. In addition, where the sapwood is completely preserved, the outer ribbed cambial surface is frequently seen in contrast to the smooth facets left by axe or adze (see Plate 2(d)).

(2) Where a series of pieces of timber from a building or structure all show the apparent outer year of their sapwood to be the same year, it can be inferred that this indicates the date of felling of all of the trees in the group.

In addition to the mechanical damage mentioned above, the sapwood of oak is subject to damage by decay and attack from wood-boring insects. Often samples are obtained with the sapwood either wholly or partially destroyed. There are three possible conditions:

(1) part of the sapwood is present;
(2) none of the sapwood is present but it is possible to ascertain with certainty the heartwood/sapwood transition;
(3) the sapwood and part of the heartwood are both missing.

In the first two cases it is possible to obtain an approximate felling date for the tree by adding an estimated number of sapwood rings to the date of the last heartwood ring. Estimates have been worked out in Germany by Hollstein (1979) and Huber (1967), but it is apparent that these are not universally applicable (see below). Hollstein has suggested figures of 20.4 ± 6.2 for 100-200-year-old trees and 25.9 ± 7.5 for trees over 200 years in western Germany. Huber seems to have varied between figures of 20 and 25. A figure worked out at Belfast suggests 32 ± 9, which is noticeably larger. This discrepancy is discussed in more detail below.[4]

In cases (1) and (2) above, an estimate of the felling date of a tree can be obtained by adding an appropriate mean sapwood value to the mean outer year of the heartwood. The use of the mean outer year of the heartwood is essential in cases where the outside of the heartwood is irregular. This method of estimation can be used with success in cases where a number of timbers from a group all exhibit some sapwood. It is then possible to ascertain, within limits, whether or not the group is of the same date. Obviously, the presence of total sapwood on a single sample from a group will always form more significant evidence than any number of samples with partial sapwood.

The most difficult case to deal with is (3), where the sapwood and the heartwood/sapwood transition are both absent. The Frontispiece shows three samples in this category, all from different buildings. If these samples were the only evidence available it would be impossible to estimate the relative felling dates of the trees used in the three buildings, since in each case a completely unknown number of rings is missing. It cannot be assumed that the existing outer ring of a heartwood beam will be close to the heartwood/sapwood transition. Sometimes beams were cut from the centre of the heartwood and in one observed case, QUB 542 from Hillsborough Fort (Chapter 5), 170 years were missing. Obviously the estimation of the felling date of a single sample with no heartwood/sapwood transition is hazardous. However, where a large population of samples is available from a single building it is possible to estimate the felling date even when no individual sample exhibits a clear heartwood/sapwood transition. The principle behind this method of estimation is given below.

In the production of oak beams two basic methods can be employed, either radial splitting or sawing. Using the first of these methods, each beam produced will contain a complete ring record from the tree centre to the sapwood. If a group of oak timbers was originally produced by splitting, then even allowing for sapwood loss and decay, a significant proportion of the group will retain their ring records to a point close to the heartwood/sapwood transition.[5] Where beams have been produced by sawing, the proportion of timbers cut entirely from the heartwood increases with the size of tree and the number of beams cut from it. Where 2 or 4 beams are cut from a single trunk, each beam will approach the original outside of the tree at some point. However, when as many as 12 to 16 beams are being cut from a large tree, between one-quarter and one-third of these will come from the centre of the heartwood and their ring records will end considerably before the heartwood/sapwood transition. In either of these two methods of oak beam production a large proportion of the timbers originally retained some trace of the outside of their parent tree. The method of estimation of felling date, outlined below, for groups of timbers without sapwood depends upon this supposition.

If a group of timbers from one building is cross-correlated, then their ring patterns can be set in relative position against a time-scale. The resulting histogram can take one of a number of forms. Figure 2.2(a) shows a case where the timbers in a group were produced by radial splitting. Originally these beams all extended to the outside of the tree. If only their sapwood is missing the number of missing rings should be determined randomly around the mean sapwood number; hence the distribution of the existing outer rings should be linear. Figure 2.2(b) shows a similar representation for a group of timbers originally produced by sawing. In this case, while the samples which originally approached the outside of the tree retain a linear distribution, the outer years of samples cut from the heartwood have many more years missing and the distribution of outer years tails off. Estimation of the felling date for a group depends on the distribution of the outer rings of the existing samples being linear. It is reasonable to suggest that in each case only the sapwood is missing and the best estimate of the felling date of the group would be from the mid-point of the linear group. One corollary of this is where, within the tail-back, a secondary linear grouping occurs. This could be strongly suggestive of an earlier felling phase even in the total absence of sapwood (see Figure 2.2(c)). Reference to Figure 8.2(c) shows such a grouping in the earlier phase of timbers from Glasgow Cathedral.

One obvious question relates to the degree of linearity; what classes as a linear distribution in this context? The answer has to be largely empirical. The spread of the existing outer rings should reflect the spread expected in the sapwood numbers. A useful working figure would be something less than twenty years for a group of at least five samples. Even had no trace of sapwood existed, the

**Figure 2.2: (a) Expected distribution of outer heartwood years in a group with sapwood only missing. (b) Expected distribution with some heartwood samples present. (c) Recognition of possible multiple phasing in the distribution of outer heartwood years.**

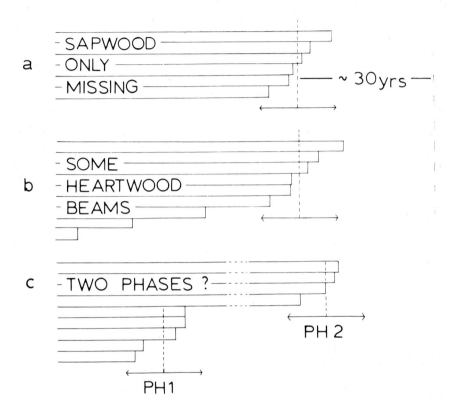

configuration in Figure 8.2(c) would under this criterion have allowed suggestion of the earlier 1258 phase. In practice the sapwood estimate would have to be added to the mid-point of the linear distribution; in this case this would have given an estimated date for the earlier phase in the range 1254 ± 9. This line is pursued further in Chapter 8.

*Sapwood Estimates*

As mentioned above, two estimates exist for the number of sapwood rings in German oaks. In 1973 measurements were made on a series of modern and post-medieval Irish oaks which exhibited total sapwood. Thirty-seven samples were measured. In each case measurement was made of the mean number of sapwood rings, the age of the tree and the mean radius. On the basis of these measurements two distributions were plotted. The first was a plot of the number of sapwood rings against total age of sample. The second represented the number of sapwood rings against the average number of rings per centimetre for each sample. In both cases the modern samples are denoted by crosses.

The first of these distributions, Figure 2.3(a), shows that for trees containing more than 80 growth rings, the number of sapwood rings is largely unrelated to the age of the tree.[6] This indicates that the parenchyma cells in mature oaks have a lifetime of between 20 and 50 years and that the heartwood/sapwood boundary advances by approximately one ring per year in middle age. The second comparison, of number of sapwood rings against rings per centimetre, indicates that the number of sapwood rings is largely independent of the width of the rings in a tree. However, it is clear from Figure 2.3(b) that the modern samples contain less rings per centimetre than the post-medieval timbers. This tendency for modern oaks to exhibit wider rings is presumably a reflection of their unrestricted growth conditions compared with the earlier forest timbers. It may also reflect the fact that the modern trees were consistently sampled low on the stem. The most important result of these distributions is the apparent randomness of the number of sapwood rings, regardless of ring width, in trees containing more than 80 rings. Thus a mean sapwood value should be valid for all mature trees.

The arithmetic mean of the number of sapwood rings from the 37 Irish samples was 31.8. Calculation of one standard deviation suggested that a useful figure for most Irish oaks would be 32 ± 9. This estimate has been used consistently since the early 1970s at Belfast. Obviously this number is significantly greater than either of the German estimates quoted above. However, it does not appear to be a case of one estimate being incorrect. Since 1973 one further sapwood estimate has been published. This was for a sample of 38 modern oaks from Maentwrog, Wales (Leggett *et al.*, 1978, 192). These trees were sampled at 'mean breast height' and yielded a mean of 27.2 rings with a standard error of 0.8. More recently further detailed study of the Maentwrog trees has shown that the number of sapwood rings increases with height up the stem and it has been suggested that a round figure allowance of 34 ± 7 rings would probably be more applicable to building timbers (where sampling height is unknown) (Milsom, 1979, 72).

A different approach has been suggested by Fletcher. This relies on the assumption that the actual width of sapwood on mature oaks is relatively constant. Fletcher suggests about one inch. Therefore, depending on the width of the final (remaining) heartwood rings, it might be possible to estimate the number of sapwood rings, for example if the outer rings averaged 1 mm there should have been around 25 sapwood rings. Unfortunately, this assumes that

Figure 2.3: (a) Number of sapwood rings against tree age (Irish oaks). (b) Number of sapwood rings against average ring width (Irish oaks). (c) Frequency of sapwood numbers in a further sample of 28 modern Irish oaks. For comparison see Figure 12.5(b).

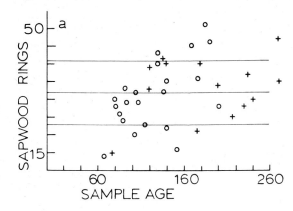

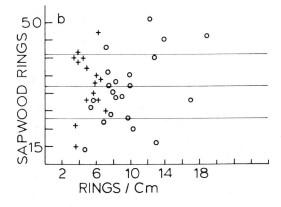

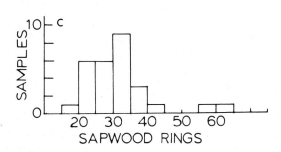

there will not be any rapid short-term changes in ring width. Using this approach to sapwood estimation, in conjunction with panel paintings of known date, he has shown an estimated mean close to 20 years and a maximum possible variation of between 20 and 30 rings (Fletcher, 1978b, 303).

The difference in sapwood estimates seems to divide between Germany — Oxford and Belfast — Liverpool. The implication has to be that these differences are real and that parenchyma cells live longer in oaks in western Britain and Ireland. It is also clear that the number of sapwood rings can be very diverse. At Belfast the samples used in the mean had from 12 to 50 rings. In fact, extreme examples have since been found with sapwood containing up to 90 rings. At the other extreme some samples have been found locally to exhibit heartwood out to the bark (see Plate 1(e)). Fortunately these extremes are rare, but it might reasonably be asked why the Belfast estimate has not been updated.

The answer to this lies in the nature of the whole sapwood problem. If the sapwood is missing, it is dangerous to assume that some magical estimate will restore the accuracy of dendrochronology. The unfortunate truth is that it will not. Whether for Ireland the best sapwood estimate were 32 or 27 would not fundamentally change the situation that in any given case the chances are that the number of sapwood rings will be somewhere between 20 and 40. To give an example, two of the modern sites cored in 1978 (see Chapter 4), Cappoquin and Glen of Downs, were analysed. These provided sapwood numbers for 28 trees cored at mean breast height. Figure 2.3(c) shows the number of samples for any given sapwood interval. Eighty per cent lay within the limits specified by $32 \pm 9$. So the estimate seems fair and no more precise figure would have significantly improved the picture. The extreme outliers could not be accounted for by any of the known estimates and it has to be accepted that in estimating the felling date of a single sample with missing sapwood the possibility exists that even $32 \pm 9$ might give a considerable underestimate of the actual felling date (Reference to Plate 1(e) shows the other extreme, where heartwood can exist locally right to the outside of the tree.)

It is important that in any estimation of felling date the sapwood allowance used should be quoted. It is, after all, possible that some better system may ultimately be available.

## Dating Quality

One obvious follow-on from sapwood estimation is the development of a scheme for dating quality. It would be dangerous to assume that, because dendrochronology can date the outer growth rings of samples to specific years, all tree-ring dates are accurate. Clearly, with incomplete samples the last growth ring may be accurately dated, but the felling date can only be estimated. Therefore complete samples must provide better dates than incomplete samples. The scheme in Table 2.1 attempts to grade the relative qualities of dating. If this or some equivalent scheme were adopted, then a grading could accompany each tree-ring date. This would allow the users of dates to assess at a glance the real quality of the date.

**Table 2.1: Suggested Dating Qualities**

| | |
|---|---|
| Type A (precise) | where the final growth ring is clearly present and either bark or 'waney' edge signifies felling. |
| Type A1 (precise) | where felling ring is identifiable as in A but where only the spring vessels of the final growth year occur. This may be suggestive of felling in the early summer, but might also in some cases indicate defoliation in the final growth year (see Plate 2(e)). |
| Type B (close estimate) | where sapwood is mostly complete but the outer portion is missing or damaged. Here the date is derived by adding an estimate of the likely number of sapwood rings to the date of the heartwood/sapwood transition. However, the felling date must be after the date of the outermost surviving ring. |
| Type C (reasonable estimate) | where only a trace of sapwood remains or where clear evidence of the heartwood/sapwood boundary can be deduced from the curvature of the surface. The felling date is estimated by adding a sapwood allowance to the date of the heartwood/sapwood transition. |
| Type D (suspect) | where no trace of sapwood exists nor any evidence for the curved heartwood/sapwood transition. In this case it is impossible to be sure of how many heartwood rings are missing, since the structure of inner and outer heartwood is identical. The date of felling will be at least $(32 \pm 9)$ years after the outermost surviving ring but may be *much* later. Therefore Type D dating supplies only a *terminus post quem* for felling. |

There is an immediate caveat to the whole question of felling years. The dendrochronologist can identify in some instances the complete, final growth ring. It is easy to fall into the habit of referring to this final growth ring as the 'felling year' of the tree. In reality this need not be strictly correct. If an oak has completed its growth for a particular year, then it is impossible to deduce whether the tree was actually felled later that same year or at some time before the onset of growth in the next year. Very often, therefore, the actual use of the timber may relate to the year after the final year of growth. The situation is less difficult where Type A1 is concerned, as the presence of spring vessels alone normally implies summer felling in the final growth year.

Plate 2: (a) An 'impossible' band of rings, where the individual lines of vessels cannot be resolved with any degree of certainty.

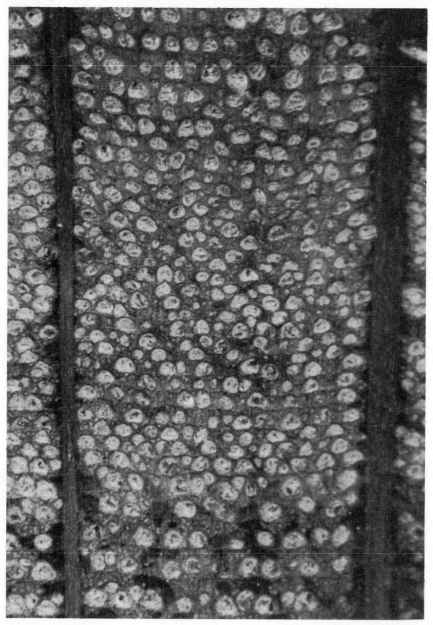

Plate 2: (b) A gross example of rings staggered across a ray. Note how the out-of-phase rings look consistent.

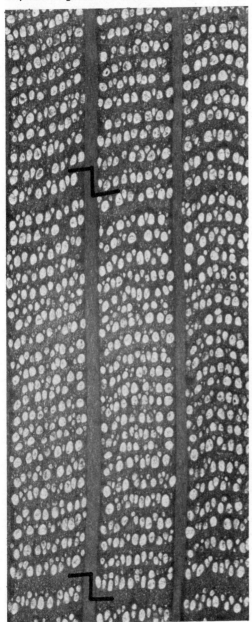

Plate 2: (c) Dramatic curtailment of growth in the year AD 1734 possibly associated in this localised instance with severe May frost.

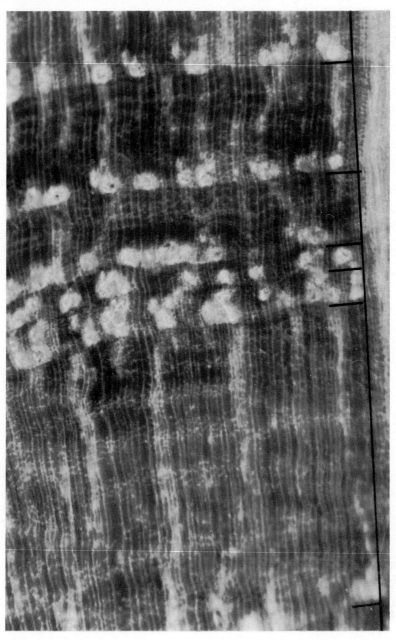

Plate 2: (d) Clear contrast between a complete underbark surface and the facets resulting from light axe or adze trimming.

Plate 2: (e) Heartwood/sapwood distinction showing clear colour difference. In this example from Gloverstown, Co. Antrim, the final year, 1716, is represented by spring vessels only, indicating felling in the early summer.

Plate 2: (f) Charred plank from Carrickfergus, Co. Antrim, successfully dated to AD 1551.

## Notes

1. What follows is no more than a basic introduction to growth in oaks. For the reader interested in knowing more about growth generally and oak in particular, the following two books are recommended; *Tree-Rings and Climate* (Fritts, 1976) and *The British Oak* (Morris and Perring, 1974).

2. The reason why we do not simply measure the widths of the summer-wood is explained by the difficulty in assessing where the summer-wood starts in many cases (whereas the start of the year's growth is normally clear). Also, since the width of the spring vessels is fairly constant, their inclusion probably causes less disturbance to the ring pattern than the rather subjective measurement of summer-wood alone. For dating purposes this poses no problems. However, for climatic reconstruction the inclusion of the spring-wood may result in the carry-over of information from the previous year.

3. The possible absence of rings may explain why some otherwise perfectly ordinary samples fail to cross-date. A missing ring would have the effect of randomising part of the ring pattern (especially serious if it occurs near the middle). Checking the pattern on the wood itself would simply confirm that the observed rings were correct and the sample would ultimately have to be abandoned. However, the fact that only a small percentage of timbers fail to date suggests that the problem is not a serious one.

4. Hollstein also quotes figures for oaks in Anatolia — $20.6 \pm 4.1$ for 100-200-year trees and $29.4 \pm 7.1$ for trees over 200 years.

5. All that is likely to be missing in these cases is the eroded sapwood.

6. Little consideration needs to be given to trees with less than 80 rings, as in most cases these would be of no use for dendrochronological dating. The problem with short patterns is that they can be 'dated' in a number of places — therefore they cannot be dated at all!

# Chapter 3

# The Practicalities of Dendrochronology

The first step in any tree-ring exercise must be the acquisition of suitable timber samples. The attitude taken towards these samples should be governed by two main factors. The first relates to the quality of the timber being sampled, as this determines the level of destructiveness which is permissible. The second relates to the true purpose of sampling, which is the acquisition by the dendrochronologist of a set of ring-width measurements. This latter point is stressed because in the final analysis the ring pattern of the timber is all that is necessary for dendrochronology. The dendrochronologist's aim should be the most complete possible ring pattern. A clash of interest can arise here because the dendrochronologist, in order to establish the best possible ring pattern, may need to be destructive. Conversely, sampling to be as non-destructive as possible may lead to an inadequate ring pattern.

Of course pieces of timber come in a wide variety of forms and conditions. These could be listed as modern, historical, art-historical, archaeological and subfossil. With the exception of modern, living-tree specimens the condition of the timbers may vary from fresh to dry-seasoned to dry-rotted to water-saturated-sound to water-saturated-deteriorated to charred to fragmentary. The shape of each sample could vary from a complete disc to a section of a beam to a radial wedge or a narrow radial plank. In general sampling includes, in decreasing order of destructiveness, cutting (slicing or half slicing), coring and edge cleaning for *in situ* measurement or photography.

Plate 3(a) shows a Swedish Increment Corer which can be used for the sampling of living trees. While ideal for conifers, it can be used successfully to core living oaks despite their higher wood density. The trick for using these corers in oak is to back off the corer half a turn for every one or two turns into the tree. Using this technique, oak cores up to 350 mm length can be obtained. In use the corer is kept as nearly horizontal as possible and aimed at the centre of the tree. Once the corer has been screwed in to maximum or required depth, a thin metal 'spoon' is inserted down the inside of the corer between the metal wall and the core. This action serves to jam the core in the corer so that further backing off snaps the core at the inner end. Backward pointing teeth on the end of the spoon allow the core to be withdrawn.

**Plate 3: (a) Working end of a hand-powered Swedish increment corer with the resultant core mounted for measurement.**

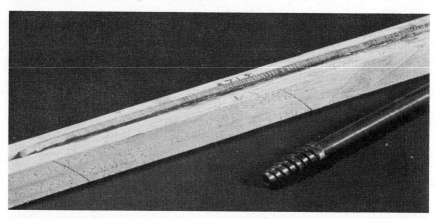

**Plate 3: (b) Mechanically powered Henson dry-wood corer and mounted core.**

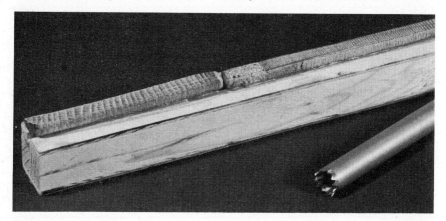

Plate 3(b) shows a Fred Henson dry-wood corer. This implement mills an annular hole, leaving a solid core of wood in the centre. Once optimum depth has been attained the corer is withdrawn, leaving the core still attached at the inner end. A thin L-shaped probe is then inserted down the side of the core and twisted to break off the core, which can then be pulled free. The major drawback of this exceedingly useful tool is its inability to self-clear, i.e. the dust product of the milling teeth has to be removed by continually pumping the drill in and out as the hole is being cut. It does, however, allow the successful sampling of timbers from standing buildings and some of the more robust museum objects.

In general, complete slices form the best samples since it is possible to choose an optimum radius for measurement. In addition, slices are more likely to retain sapwood. This is an important point, since the precision of the final date depends upon the degree of sapwood completeness. If dry-coring is undertaken it is almost certain that any sapwood will be destroyed by the drill action. It is necessary, therefore, in cases where dry-coring is used, to acquire additional sub-samples of sapwood from the same timber whenever possible. With water-saturated samples, where the wood has lost most of its substance, some workers have had success with freezing. Frequently measurement while the sample is still wet is used, especially with sapwood, which is likely to deteriorate badly if left to dry out.

Overall, the approach regardless of condition is to extract the ring pattern by any means possible. In theory the sample could then be discarded, though in practice it is kept for reference purposes. Not nearly enough use has been made of photography for recording ring patterns. For example, it would almost certainly be easier to photograph the cleaned edges of panels rather than attempt measurement *in situ*. The permanent photographic record could then be measured under a microscope, a suitable scale being achieved by marking points on the sample a known distance apart. The author has successfully dated a wooden panel from measurements made on a 35 mm transparency of the original. Ultimately it may be possible to sample solid wooden objects, for example medieval sculpted figures, with a variant of the medical 'whole body' scanner, the actual measurements being made on an X-ray plate. In the same vein ultrasonics or microprobes may become available in suitable forms. In the meantime, in the absence of X-ray spectacles a useful motto would be 'Cut and be damned.'

## Sample Preparation

As with sample acquisition, preparation of the sample for measurement is aimed solely at producing a reliable ring pattern. Thus any procedure which clarifies the ring record, be it sanding, planing or paring (with a razor or scalpel), should be acceptable. An extension of this to staining the wood or highlighting the spring vessels with chalk or some other medium only makes common sense. The simplest samples to deal with are dry building timbers. Oak is generally robust and decay resistant and the heartwood can be brought to a high polish with a rotary or band-sander and fine emery. Plate 4(c) shows the stages of preparation of a seventeenth-century sample. The final polished section has the spring vessels highlighted with chalk. In this sample the sapwood is missing. Where sapwood remains on dry building timbers it has normally suffered considerable insect damage. Plate 1(d) shows an example from Springhill (see also Chapter 6). In such cases the sapwood can be prepared by fine sanding, the problem being to find a radial section sufficiently intact to measure.

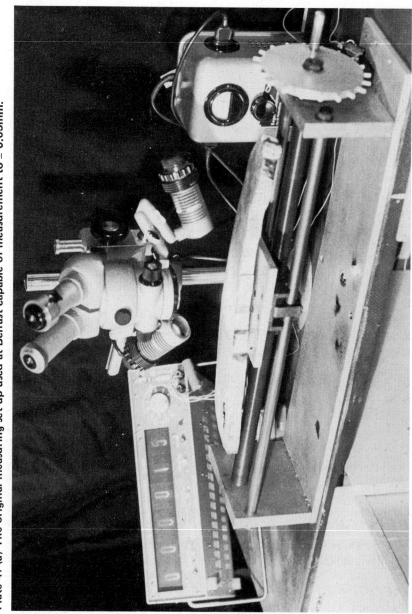

Plate 4: (a) The original measuring set-up used at Belfast capable of measurement to ± 0.05mm.

Plate 4: (b) More recent Henson-made 'Bannister' measuring device using linear transducer and automatic data print-out.

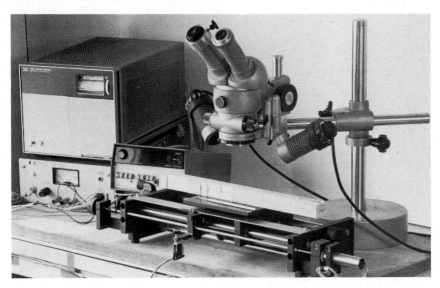

From an archaeological viewpoint many more samples are encountered wet. While the heartwood will normally retain much of its substance and can be dried out slowly and polished as above, the sapwood must be treated rather differently. Wet sapwood on ancient samples will seldom survive drying out. Unlike the heartwood, which was consolidated before the tree was felled, the sapwood will have lost most of its soluble components, leaving only a tracery of cell walls. In drying, these remaining structures collapse, leaving no more than a wafery residue. There are two solutions. Either the sapwood can be measured wet and the sample then allowed to dry, or the sample can be frozen or kept wet. The important factor is the accurate recording of the ring patterns to the felling year.

With both wet and dry cores the procedure is to mount them in wooden channels as in Plates 3(a) and (b), and either pare or sand a flat transverse surface. The snag with such cores is undoubtedly the narrow radial section which is available for examination. With discs or slices it is possible to choose an optimum radius for measurement; with cores it is not. Given the radial structure of oak and the difficulty of accurately judging the position of the centre of the tree, many cores have to be measured by continually crossing rays. Measurement consists in these cases of a series of overlapping sections. Each jump across a ray requires care because of possible staggers in the lines of vessels (see also Plate 2(b)).

**Plate 4: (c) Typical dry-wood sample in stages of preparation for measurement.**

## Ring-width Measurement and Problem Rings

By convention the ring pattern of a timber is defined as the successive widths of
each of the rings from the pith (centre) to the bark. Since in practice samples are
often incomplete, this involves measuring from as close to the centre to as close
to the outside as possible.[1] Each ring is measured from the start of its spring-
wood to the end of its summer-wood. This is achieved by measuring from the
start of one row of spring-vessels to the start of the next, the measurement being
at right angles to the rows of vessels. So in theory at least the establishment of
the ring pattern of a sample is a simple procedure. There are, however, several
complicating factors which lead to rather more difficulty in practice. A few of
the problems regularly encountered are discussed below.

First, the measurement process. Most of the ring widths to be measured will lie in the range from a few tenths of a millimetre up to several millimetres. It is desirable that the individual measurements should be accurate to something of the order of 0.05 to 0.02 mm. The latter figure would seem to be more than sufficient. Tree-rings are, after all, part of a natural system with built-in variability. There would be little point in measuring to accuracies beyond 0.01 mm as the relative variation even within the same ring would be much greater than this. All the early work at Belfast was carried out on equipment which measured to an accuracy of 0.05 mm and yielded perfectly adequate results. It should be remembered in this context that the dendrochronologist is really interested in the width of each ring relative to its predecessor, as exemplified in skeleton plots in Chapter 1. This is the essential information which is being assessed when two ring patterns are compared, as will be seen when cross-correlation techniques are discussed below. So when, for example, a dendrochronologist refers to a narrow ring he is referring to its width in comparison to the previous year's growth rather than its absolute width. Dendrochronology is therefore a study of relative changes. The measurement of the ring width is simply a mathematical quantification of such study.[2]

Most equipment for measuring ring widths relies on the same basic principle. The prepared sample is placed on a travelling stage and viewed (preferably) through a binocular microscope with cross-hairs. The stage is capable of horizontal movement which can be accurately measured. So to measure the width of a ring the cross-hairs are aligned with the start of the spring vessels of the ring and the stage is then moved until the cross-hairs align with the start of the spring vessels of the following ring. The distance travelled by the stage is then read off and recorded. Now such a procedure can be carried out simply or with increasing complexity. For example, Plate 4(a) shows the early Belfast equipment. The stage was drawn along a threaded bar of 1.0 mm pitch, i.e. one turn of the threaded bar moved the stage 1.0 mm. On one end of the bar was a disc with twenty studs. As the disc was rotated the studs triggered a microswitch, registering the distance moved by the sample in units very close to 0.05 mm. More recent systems differ only in the use of linear transducers or rotary encoders to measure stage movement or threaded bar rotation respectively. Such systems normally print the ring-width measurements automatically on to paper tape. Obviously more sophisticated set-ups are possible with direct links to microprocessors. However, a cautionary word is necessary here. It is important in dendrochronology that the operative should interact strongly with the data; in particular this applies in the cases outlined below, where decision-making becomes necessary.[3] It is often desirable to annotate the data sheets in cases of doubt or with regard to ring peculiarities. Moves towards over-automation could well be counter-productive in giving an artificial integrity to a set of numbers.[4]

As the sample is being measured it is desirable to mark rings physically at set intervals, for example every tenth ring. If this is carried out systematically, it is easy to scan the sample visually and check the spacing afterwards — two of the commonest errors in dendrochronology are losing a ring where a sample has been realigned during measurement or gaining an extra ring where a dubious ring has been measured twice. Some workers prefer to mark the sample every ten years before measuring. Each batch of ten rings must then end at a marked ring — failure to do so indicates an error. The other principal advantage in systematic marking, with double marks every 50 years and treble every 100, is as an aid to the relocation of specific rings for checking or whatever.

*Some Ring Anomalies*

Oak is a ring-porous wood. This definition indicates that the spring vessels are noticeably larger than the summer vessels and that they occur in a distinct row at the beginning of the year's growth. While it is clear that as a rule oak trees put on one growth ring per year, there are cases where the interpretation of the combinations of spring vessels can be difficult. Most such difficulties arise because the rows of spring vessels can be more than one vessel wide and occasionally differentiation of the individual rows of vessels into definite rings can be doubtful, introducing uncertainties into the otherwise precise time series. A variation of the same problem arises with the differentiation of the individual rings within bands of narrow rings.

Figure 3.1(a) shows a situation where a tree has been developing a single row of vessels each year. At some point it shows what appears to be a very narrow ring — or is it one ring with a double row of vessels? Figure 3.1(b) shows the converse situation where a tree which has been developing double rows of vessels shows two distinctly separate single rows. Does this represent a narrow ring or is it simply an example of dispersed vessels within a single ring? Experience tends to suggest that in each of these cases the simple solution will tend to be correct. However, that does not stop an element of doubt being introduced. To give a concrete example: Plate 1(b) has already been cited as an example of a 'ghost' ring in the guise of a row of enlarged summer vessels. This could only be known with hindsight. When originally encountered it was not known whether this represented a peculiar diffuse ring or simply a 'ghost' ring.

So in the measurement process rings are encountered where the operative, although fairly certain, cannot be absolutely sure of the correct interpretation. Such rings must be recorded with 'question-marks' as a guide to likely points of error. Where they occur, they reduce confidence in the overall ring pattern.

Figure 3.1(c) shows a situation encountered in trees which presumably suffered a sudden, severe set-back. A tree has been developing consistent double rows of vessels when suddenly it shows a ring with no summer-wood. The

**Figure 3.1: (a) Narrow ring or double vessels? (b) Double vessels or narrow ring? (c) Catastrophic reduction in growth characterised by a change from double to single vessels (see also Plate 1(f)).**

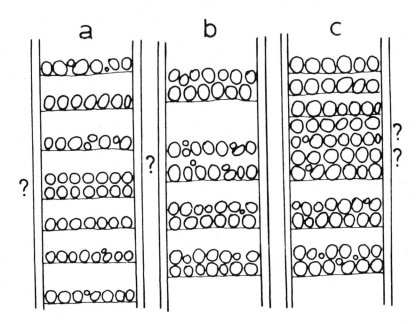

following year it develops a single row of vessels. Further single rows may be encountered for a number of years after the event until the tree regains its former 'strength' (growth potential). When first encountered this phenomenon can be difficult to interpret. Again experience suggests that the first two rows of vessels belong to the year of defoliation, the next *single* row to the year after, the next to second year after, etc. The reason for this interpretation is that the presumed defoliation takes place after the spring vessels were formed. As they were formed on the strength of reserves laid down in a 'normal' year they should be normal also, i.e. in this case there should be two rows. The defoliation results in no summer-wood formation and hence the following year reserves are low, resulting in a single row of vessels. However, this interpretation is possible only with hindsight. Plate 1(f) shows an example of this phenomenon.

One of the worst configurations to be met with is the batch of extremely narrow rings sometimes encountered within an already narrow band. Such a case is shown in Plate 2(a). What seems to happen is this. At the bottom of Plate 2(a) the tree is developing narrow rings with single rows of vessels. The rows are clear and in most instances it is possible to see a line of very small cells, normally associated with leaf fall, just before each row of vessels. At some point this clear differentiation of the individual narrow rings breaks down and we are confronted with a band of ten or twenty rings which cannot be resolved. The previous rows of vessels, which the eye could easily resolve into separate rings, are replaced with a situation where the vessels of one ring effectively overlap with those of the next. This breakdown lasts for a number of years (occasionally as many as forty), after which the tree is able to return to apparent normality. It is hard to understand the mechanism for such growth, since the cambium is normally a continuous layer. In these extremely narrow, unresoluble bands it has to be concluded that cambial activity, for whatever reasons, becomes discontinuous. One thing is clear — when this phenomenon is encountered it is impossible to measure across the band with certainty. The degree of uncertainty can be as bad as ± 5 years. Unfortunately it seems that such problems tend to be met with in the more important samples!

Other anomalies have been noted; for example, locally absent rings (see Plate 1(c)) and rings which are seriously staggered across rays. This latter can be a potential source of error, especially in the measurement of cores. Plate 2(b) shows how direct measurement, apparently across to the same row of vessels, could lead to an error of two years in a ring pattern. When moving from one radius to another it is essential that such anomalies be checked for by counting back (or forward) to obvious wide or narrow rings which are clearly consistent across the rays.

Now each of these phenomena has the effect of introducing question-marks into the basic ring-width data set. This, as will be seen below, can have serious consequences for cross-dating. Let us forget for the moment which particular anomaly has given rise to a problem and simplify the situation down. Say we have a 150-year ring pattern with one ring which in reality may not exist; the possibilities are that we indeed have a 150-year pattern or that we have a 150-year pattern with an error, i.e. with one ring too many, in reality a 149-year pattern. It would be possible to produce two ring patterns for the sample, one with the questionable ring included and one with it removed. These could both be processed and the one which cross-matched within the system could be accepted as correct, the other being discarded. However, that approach assumes that there is a 50/50 chance of the questionable ring being there or not. This is not necessarily the case. As mentioned above, experience and a certain level of expertise may well allow the situation to be not 'Is this a ring or not?' but rather 'There is a 90 per cent chance that this is a ring but still a slight chance that it is just double vessels.' In such a case we may not be justified in making two alternative ring patterns. What we have is a 150-year ring pattern with a question-mark at one ring.

This is a tricky area for the dendrochronologist and one which may be worth labouring a little further. The danger in inserting or removing rings from ring patterns becomes acute when a pattern is encountered with two or three questionable rings. If we start producing four alternative patterns (for two doubtful rings) or eight alternative patterns (for three doubtful rings) we are entering a realm where the essential integrity of the time series is being tampered with. We have to go back to the basic hypothesis behind dendrochronology, i.e. the ring pattern of wide and narrow rings which grew over a *unique* span of years and hence is unlikely to be repeated. It should therefore match in a unique position in time against a reference chronology. If we begin subdividing a time series and moving the sections until a match is found, we are no longer within the original premiss of the method.

The only solution is for the dendrochronologist to spend sufficient time on the resolution of problem cases to weight the odds in favour of each decision being correct.

## Data Representation

Having prepared and measured an oak sample, the dendrochronologist has moved away from a piece of wood to a set of numbers: a time series. The first problem is how to represent this numerical series visually. There are choices. In much of the south-western United States the obvious narrow and wide rings were in themselves sufficient to allow cross-dating. This gave rise to the use of the skeleton plot (Chapter 1). Skeleton plots convert the patterns of wide and narrow rings into what is in essence a histogram where narrow rings are represented by vertical lines, complacent rings are omitted and wide rings are designated B. An example of this technique is shown in Figure 1.1(a). The height of a vertical line reflects the degree of narrowness of the ring. This sort of representation is only of use where the trees are extremely sensitive, as in semi-arid regions. With deciduous trees, skeleton plots neglect a considerable amount of information stored in the relatively complacent rings between signatures. In studies on oak most work has been carried out using either direct ring-width plots or log ring-width plots. The use of logs has the effect of accentuating the narrow rings and damping the wide rings. As a result the log-plots show a flattened profile compared with the absolute ring-width plot. In all cases the horizontal scale is in years. In practice most German and English workers use semi-logarithmic paper and plot the log-widths against a horizontal scale of one ring per half-centimetre. The use of this elongated scale has been justified in that it can be easily annotated. In Ireland all work has been carried out using raw ring widths plotted against a horizontal scale of one ring per quarter-centimetre. The principal reason behind this separate convention is the fact that at any one time the human eye can take in detail across a field about 25 cm wide. This means that using the tighter scale (0.25 cm) it is possible to scan a century of tree-rings at one time. Using the larger (0.5 cm) scale this is reduced to about half. Since useful ring patterns normally have a length of over 100 years the tighter scale seems more practicable. The choice is of course a matter for the individual, but once made it is comparatively difficult to change.

## Cross-correlation of Tree-ring Patterns

Any dendrochronological study involves the establishment of cross-correlations between ring patterns. This is true of chronology building, dating exercises involving comparison against reference chronologies and investigations into similarities between reference chronologies from different areas. The underlying assumption is that significant correlation *must* exist between the ring patterns of timbers which grew under the same conditions over the same period of years. Further, comparisons between the ring patterns at any other relative positioning must approximate to correlations between sets of random numbers. Obviously if these two conditions are not fulfilled, dendrochronology is not going to work because it will be impossible to find definite matching positions.

First let us consider visual matching. The early American work relied almost exclusively on visual observation of the timber samples and skeleton plots. This was made easier because the key signature patterns occurred as groups of obviously wide and narrow rings. In Europe the whole ring pattern has to be considered, since visual signature matching with oak is essentially impossible. So, when the successive ring widths are plotted against a scale in years, the resultant tree-ring pattern presents itself to the eye as a random jagged curve. Superimposed upon the yearly fluctuations in ring width are trends towards wider or narrower rings, indicating respectively improving or deteriorating growth conditions. These trends can be long- or short-term and of varying magnitude. Visual comparison of ring-width plots involves superimposing the two curves under study and shifting their relative positions until such time as significant agreement is obtained between them. In practice the observer looks at significant features, such as wide or narrow rings, narrow bands, trends or notable patterns in one curve and attempts to duplicate them in the second curve. However, visual matching can be subjective. The ability of a trained observer to find sufficient similarities in two long curves to establish a cross-correlation is not a measurable quantity. Thus it is essential, for consistent results, that some repeatable measure of the significance of a cross-correlation should be independently produced to substantiate each visual match. There is an important point to be made here. It is not that individuals cannot find the correct matching positions visually; they can and often do with

considerable expertise. The problem for the observer, for example the archaeologist, is in knowing whether any particular dendrochronologist possesses the
ability, hence some mathematical quantification of each visual match is necessary.

*Statistical Methods Used in the Cross-correlation of Tree-ring Curves*

*Non-parametric.* Huber drew attention to the subjective nature of visual matching
of tree-ring curves and quoted a powerful non-parametric statistical method of
checking visual agreement. This method involves calculating the percentage
parallel variation between two ring curves. This is a measure of the number of
years where the two ring curves under comparison show similar increases of
decreases in ring width. The expected percentage agreement for two random
curves is 50 per cent, since for a long overlap there will be as many years agreeing
as disagreeing. As with any statistical results, the actual percentage agreement
figures produced in the random mis-matching of two ring curves are given by

$$\%A = 50 \pm \frac{50}{(n)^{1/2}}$$

where n is the number of years under comparison and $\frac{50}{(n)^{1/2}}$ is one standard deviation (Huber and Giertz, 1970, 203).

At any mis-match position, the correlation between two tree-ring curves
approximates closely to that between two random curves, and the percentage
agreement figures will be distributed around 50 per cent. Figures as far as three
standard deviations from the mean are likely to occur about once in every
thousand random comparisons. If we compare two random sets of numbers with
overlaps of 100 years (the shortest match likely to be definitive) then the standard
deviation will be ± 5 per cent. Thus one in a thousand such comparisons should
be as far as 15 per cent from the mean of 50 per cent. We will be looking for
significant matches to fall outside this range and since we are only interested in
positive correlations, our back-up values should be greater than 65 per cent. That
is, in order for an actual tree-ring match to be significant it should produce a
percentage-agreement figure greater than three standard deviations from the mean.

In practice, with computer availability, it is possible to slide one curve past
the other, in increments of one year, and calculate the percentage agreement at
each point of overlap. All mis-match results should be distributed within three
standard deviations of the mean, and any value occurring outside these limits is
highly significant. Eckstein and Bauch (1969) published a computer programme
which calculates the percentage agreement figures and prints them together with
their significance levels. This method of establishing cross-correlations was used
for some time during the inception of the Belfast project. The computer programme used was written using basic principles before the publication of Eckstein

and Bauch. However, it was found to be insensitive in many cases, especially with short overlaps. Its most serious drawback was the small tolerance between the percentages produced by actual agreements and those produced by chance where short overlaps were concerned.

*Parametric.* So while significant results could be obtained using the percentage agreement method, a more powerful statistical method was sought for use with Irish oak timbers. The basic problem is the matching of two sets of numbers. When mis-matched these approximate to sets of random numbers and should present low correlation figures. When significantly matched, i.e. when the curves represent the same span of years, the correlation should be high, assuming the basic model of similar growth in similar conditions.

If a set of points (x, y) show a trend when plotted, x and y can be assumed to be correlated. Figure 3.2(a) shows a high correlation between x and y in which an increase in x is associated with an increase in y and vice versa. In this case the correlation is positive. Increasing x associated with decreasing y would show negative correlation. Figure 3.2(b) represents uncorrelated values of x and y. If we consider x and y to be the ring widths of two trees growing over the same period, then, on the assumption of similar growth patterns, in year i, an increase in $x_i$ should be associated with an increase in $y_i$ or a decrease in $x_i$ with a decrease in $y_i$. When this is true, a high positive correlation should result.

So the basic assumptions of the dendrochronological method argue strongly for the use of a direct parametric correlation method. The degree of correlation between x and y is measured by 'r', the produce-moment correlation coefficient. This is defined as

$$r = \frac{\Sigma_i x_i y_i - N\bar{x}\,\bar{y}}{\sqrt{(\Sigma_i x_i{}^2 - N\bar{x}{}^2)\,(\Sigma_i y_i{}^2 - N\bar{y}{}^2)}}$$

where $\bar{x}$ and $\bar{y}$ are the means of all the x and all the y values respectively. The term on the bottom line is the product of the standard deviations of all the x and y values. For 'r' to be valid, the values x and y should be evenly distributed about the means $\bar{x}$ and $\bar{y}$. Because of the overall trends towards wider or narrower rings, the simple mean of all the ring widths in a tree-ring curve will not necessarily form a valid mean at any given point, i.e. the curve may fluctuate from a simple mean value. In order to eliminate the possibility of decreased sensitivity when correlating tree-ring curves by this method, the basic data has any trends removed. In practice each ring width in the primary data is converted to a percentage of the mean-of-five ring widths. This reduces a ring-width curve to a percentage average curve about a mean value of 100 per cent. This means that we can now compare the high-frequency, year-to-year detail, i.e. the matching component.

**Figure 3.2: (a) Highly correlated X and Y. (b) X and Y uncorrelated. (c) The effects of meaning wide and narrow ring patterns. The induced feature m-n and the trend p-q are clearly not present in the individual ring patterns.**

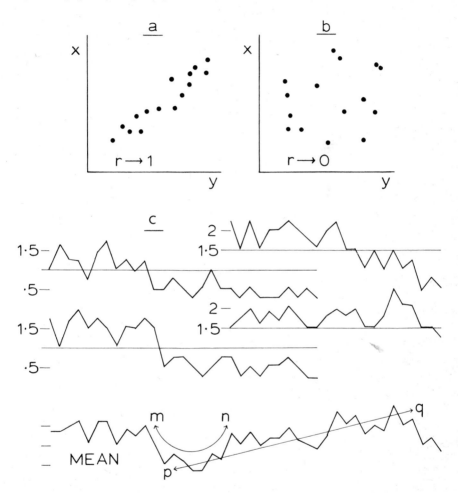

The correlation coefficient 'r' can have values between -1 and +1. For perfect positive correlation 'r' should equal 1, while correlation of two mis-matched curves should produce a value of 'r' around zero. The value of 'r' takes no account of the number of variables N. Thus if two curves are moved past one another and the value of 'r' calculated for each position of overlap (as above), the values of 'r' are not immediately interpretable in terms of probability of occurrence.

In order to relate 'r' values to probabilities the simple conversion to Student's '*t*' is used, where

$$t = \frac{r\sqrt{N-2}}{\sqrt{1-r^2}}$$

the value of *t* gives a measure of the probability of the observed value of 'r' having arisen by chance in a sample N. In practice, to approximate normality of the distribution of *t* values, the percentage-average figures can be reduced to log values. Tables exist which relate the value of *t* to probability for any value of N.

Since most useful ring patterns contain 100 or more rings, the 0.1 per cent significance level of *t* is *t* = 3.5, i.e. a value of this magnitude should arise by chance about once in every 1,000 mis-matches. This is equivalent to the three standard deviation level for percentage agreements (above). In practice the background *t* values fall between *t* = 0.0 and *t* = 3.5, and correlations of ring patterns which grew over the same span of years normally produce *t* values greater than *t* = 3.5. No account is taken of negative correlations, since dendrochronology by definition is dealing with positive similarities between ring patterns. A computer programme which calculates '*t*' values for each position of overlap between two ring patterns is available and is widely used (Baillie and Pilcher, 1973, 7-14).

At the time of publication of the Belfast CROS programme it was recognised by the authors that the *t* statistic calculated by CROS was not a true Student's *t*. As pointed out by Barefoot *et al.* (1978): 'the statistic as computed . . . has a slight error and is not a true *t*'. However, the programme was devised to look for the best possible fit of the high-frequency data in the tree-ring series, i.e. the year-to-year variations. It does not take account of autocorrelation in the ring series, i.e. the tendency in tree-rings for each ring to be influenced to some extent by its predecessor or predecessors. However, the shortcomings of CROS statistically, apart from being recognised in the publication title 'A Simple Cross-dating Program for Tree-ring Research', are more than made up for by its ability to pick up matches between ring patterns. Workers at Nottingham have suggested that the CROS *t* values should be named 'maximised *t*', since they found it impossible to make the programme more statistically robust without reducing the values produced (Triggs, personal communication). Since the dendrochronologist wants the matching correlation figure to stand out as much as possible from the random figures, the maximised *t* would seem most useful. In practice it has been found that CROS will indicate any correlation positions which are worth checking visually. A genuine match will normally yield at least *t* = 3.5, usually much higher. However, on occasions lower values need to be checked, especially when comparing patterns over long distances. Figure 4.5 shows the output produced when a 130-year ring pattern was run against a 970-year reference chronology using CROS.

In practice all computer values, percentage agreement or $t$ form a guide to others that a proposed match has at least some correlation. Of course just because a match has a value of $t > 3.5$ does not mean that it has to be correct. The final decision must always rest with the dendrochronologist – the computer is only a back-up. To restate this last and important point: because a significant $t$ value is quoted does not mean that the match is definitely correct – that rests in the hands of the dendrochronologist and the ultimate decision on correctness is his decision. However, a dendrochronologist's suggested match, if *not* backed up by a significant computer correlation, may well be suspect!

Thus dendrochronologists have several lines of approach to establishing cross-matches. The first and most powerful is the worker's own ability at visual matching and this can be backed up by the second, the use of statistical computer programmes. It is advisable that the dendrochronologist should always have a final say in acceptance of cross-matches. It is a dangerous course to allow statistics to overrule human judgement. The routines outlined above are, after all, only mathematical approximations of the resolving power of the human eye and brain. Humans have had millions of years of evolutionary experience in pattern recognition and it would be difficult to programme a computer to duplicate adequately this experience. The importance of the mathematical approximations lies in their independence.[5]

## Replication

There is a third and very important back-up to visual matching and statistics, which comes under the heading of replication. There are several levels of replication to aid the dendrochronologist. First, there is the replication between ring patterns of individual trees. This ensures that there are no problems with incorrect ring counts and, when masters are produced, reduces the 'noise' within the individual ring patterns enhancing the climatic 'signal' – the matching component. Second – there is replication between master chronologies. Each time a series of timbers is cross-dated and a master produced this is matched against a reference chronology. This process allows the checking of the relevant portion of the reference chronology and over a period of time all portions of a reference chronology will be duplicated. In short, as the dendrochronologist dates timbers he replicates more and more of his work, giving a continual cross-check. An increasingly important factor is that other workers, some in areas comparatively close at hand, will generate tree-ring chronologies which may well replicate existing chronologies. Recent evidence suggests that within Britain and Ireland as a whole, replication is to be expected at significant statistical levels (Baillie, 1978a). All chronologies will ultimately be checked by this means.

### Representation of Accumulated Data

The representation of primary tree-ring data has been discussed above. A more important aspect of representation is the treatment of accumulated data.

When a number of ring patterns have been cross-correlated the information relevant to any individual year is best represented by the sum of the information supplied by all the trees growing in that year, i.e. if a significant majority of trees show an increased or decreased width in a particular year then this dominant trend should be recorded in a master chronology covering that year. Because of this better correlation values are normally obtained between an individual ring pattern and a master chronology than between two individual ring patterns. Master chronologies tend to iron out the inconsistencies found in individual ring patterns, i.e. they reduce the noise and enhance the common signal.

There are, however, a number of different ways of representing accumulated data and it would be fair to say that no one method is ideal. In principle the dendrochronologist is trying to reconstruct the past signal to which the tree responded. The closest he can come to this signal is in the common response of a number of trees. One of the simplest approaches is to take the mean value over a number of trees, i.e. sum all the ring widths for a particular year and divide by the number of trees. This figure is broadly related to the total wood yield in a given year. This sort of master chronology is widely used and provided that the information from a sufficiently large number of trees is included, the master will be highly valid. Most qualms about mean masters relate to the question, 'What happens when you mean wide-ringed with narrow-ringed trees, won't the wide-ringed patterns swamp the narrow-ringed examples?' This is a very reasonable question when small numbers of trees are included. One problem can arise when trying to compare two mean chronologies. It is not really known whether trends in the data are real — they may reflect essentially irrelevant information. For example, Figure 3.2(c) shows schematically a possible situation where the outer ends of long ring patterns of one phase are being meaned with the start years of a later phase. The resultant mean chronology would contain a 'step feature' — a trend which was not a part of the original signal but which is an artifact of the age trends in the trees. If some of the observations on the non-continuum of long-lived trees are correct (see Chapter 11), then such unacceptable steps in chronologies could take place at regular intervals.

Now the way to get around this type of problem is to convert the raw ring-width measurements into indices of growth. There is nothing frightening about this process; in essence it involves removing trends in the data to a greater or lesser extent before the individual ring patterns are meaned. An indexing routine produces transformed ring widths which are compatible, i.e. each tree has equal weight in the manufacture of a master chronology. This seems to be the best

Figure 3.3: (a) Straight line fit to remove growth trend. (b) Fitting a tight or high-pass filter effectively removes all trends. (c) Fitting an exponential curve removes the growth trend normally associated with the ring patterns of conifers. (d) Fitting a low-order polynomial removes the long-term trends most often encountered in the ring patterns of oak trees.

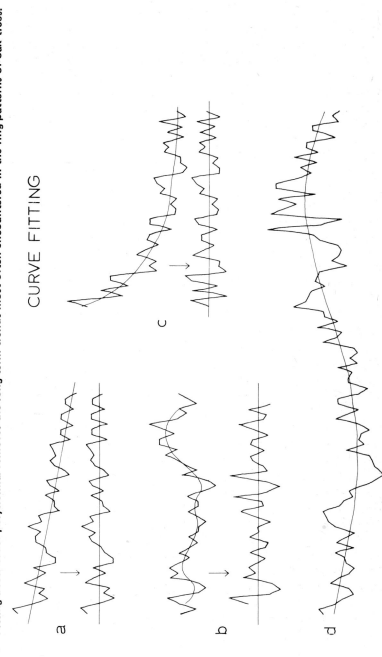

CURVE FITTING

way to proceed. In practice, some sort of curve is fitted to the data and the individual ring widths are then related to the fitted curve. Figure 3.3(a) shows a very simple example. The original ring pattern slopes slightly from wide to narrow rings. Obviously that information is irrelevant to the master, since it is characteristic of this individual tree. So a straight line is fitted to the data — a simple mathematical procedure — and each ring width is divided by the value supplied by the line at that point. Plotting the resultant indices, which fluctuate around unity, produces the flattened ring pattern shown. All of the year-to-year detail and short- to medium-term trends are preserved, with only the growth trend removed.

The question does arise as to how much trend to remove. In the example above only the overall trend was removed. It is possible to go to the other extreme and remove all trends, leaving only the year-to-year detail. This can be achieved by fitting a very tight curve to the data, as in Figure 3.3(b). A simple procedure would be to fit a running mean to the data — for example, convert each ring width into a percentage of the mean of the five rings of which it is the central value. Such a curve could be called a '%5 Average'. If this were applied to each ring pattern the resultant %5 Average curves could be meaned directly. This is the filter used to transform the original data in the CROS programme above. Figures 3.3(c) and (d) show the fitting of exponential and polynomial curves to appropriate sets of data. In general terms the exponential fit is most useful with conifers since their growth profile is normally of this shape. The polynomial fit seems to be most useful for the growth profiles encountered with oaks. The Tucson INDXA computer package enables the dendrochronologist to choose the most appropriate curve fit for each individual ring pattern (Graybill, 1979).

Two questions remain. Is it best to produce straightforward mean chronologies or indices, and in the case of indices, how much trend should be removed? There is no simple answer. Individual preference certainly comes into it, but mostly it depends what purpose the data is to be used for. For example, if straightforward dating is the objective, then anything which enhances the year-to-year detail should be an advantage. An extreme case of this approach has been used by Hollstein, where each ring width is expressed as a function of the width of the previous ring, so that a set of ring widths $X_1 - X_i$ is converted to a set of indices $Y_1 - Y_{i-1}$ where $Y_n = fn \dfrac{X_n}{X_{n-1}}$. The figures are cumulative, i.e. in order to plot out such a set of data one has to make the second ring some proportion of the width of the first ring and so on. Unfortunately this can produce a master chronology which is visually untidy, as long-term trends can be induced simply because each ring tends to be slightly narrower than its predecessor. In his book Hollstein (1979) lists all his data in this format.

For the purposes of climatic reconstruction, where the short- or medium-term trends are likely to be meaningful, removal of the growth trends from the individual ring patterns will normally be desirable and hence the various curve-fitting options would seem to offer the best approach. The trends which appear in index chronologies must be real, since they consistently have to survive the curve-fitting routine in a majority of cases to come through to the master. The trends in mean chronologies, as mentioned above, tend to be less well quantified. So it is up to the individual which formats and procedures are used, the sole proviso being that the original (raw ring-width) data should be preserved on the off chance (a) that some better system comes along or (b) that someone finds something else to do with the data, the better to extract information of, for example, climatic character.

There is one significant observation to be made on the whole question of master chronologies. It transpires that tree-rings constitute a robust system. No matter how a master chronology is constructed, from a group of definitively matched ring patterns, the similarities are striking. Figure 3.4 shows a section of the Dublin chronology (see Chapter 7) constructed by (a) using a %5 Average high pass filter, (b) meaning the raw ring widths, (c) using an INDXA polynomial option and (d) by trends (see below). It is immediately apparent that the year-to-year detail is immaculately preserved by any of the methods. The trends tend to differ. Since cross-dating is carried out on the year-to-year detail, any of the methods would appear to be useful.

Having said that, it is probably worth mentioning the simplest and quickest method of master production. This is the majority trend method shown in Figure 3.4(d). This method is useful late at night, when the computer is down and you have half a dozen matched ring patterns for which you need a working master. First align the plotted ring patterns in their correct temporal positions, then for each year count up the number of trees which put on wider rings (than in the previous year). Similarly count up the number with narrower rings. If a tree shows a ring width identical to the previous year, ignore it. So you end up with two numbers for each year, these are tabulated as follows:

| Year | 1 | 2 | 3 | 4 | 5 | 6 | 7 | |
|---|---|---|---|---|---|---|---|---|
| No Up | 2 | 5 | 6 | 3 | 0 | 2 | 2 | etc. |
| No Down | 4 | 1 | 0 | 3 | 6 | 4 | 4 | |

The resultant master is then plotted cumulatively. An arbitrary number, say 100, is plotted. Year 1 had a majority of narrow rings so plot 100 *minus* the difference, i.e. 98. Year 2 had a majority of 4 wider rings, plot 102, etc.[6] The resultant master may wander to some extent, as noted by Hollstein with his cumulative technique, but it is perfectly adequate for rough work. A master of, say, 10 trees covering a century can be generated in about five minutes.

The reason why something so simple gives an adequate master is due to the

**Figure 3.4: Section of the Dublin medieval chronology constructed using the same data in differing formats: (a) %5 Average (b) mean (c) index (d) majority trend. Note the manner in which the year-to-year detail is preserved in each case.**

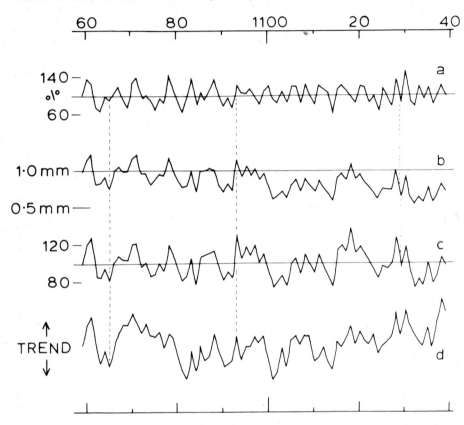

fact that it plots the majority trend, something that any technique will do. One advantage of this type of procedure is that it allows the signature years to be picked out immediately. For example, years 3 and 5 above are 100 per cent signatures. Moreover, any long upward trends in the overall master should be of interest climatologically, since such an event indicates that a majority of trees are producing wider rings over a period of years.

*Signatures*

The question of signatures is one which requires clarification. Obviously a signature is something consistent, i.e. it shows up in most ring patterns from a given area. Equally obviously, a signature cannot be deduced from a single ring pattern, since it is impossible to tell if it is consistent until compared with other trees. Therefore signatures are a by-product of the production of masters. Unfortunately signatures mean different things to different people. To American workers a signature was normally a pattern of rings which occurred clearly in most trees. The obvious matching of the bands of rings in the Frontispiece is an example. However, while such signature patterns are occasionally recognised in oak studies, for example Huber's 'saw' in the sixteenth century (Huber and Grietz, 1970), they are not found in sufficient numbers or with a sufficient consistency to act as definitive dating criteria.

On the other hand, in Europe a different type of signature is noted. This is the signature year. Huber suggested that a signature year was one where 75 per cent or more of all trees showed a similar trend to a wide or narrow ring in a particular year. It is common practice with some workers to note signature years in the main master chronologies with a heavier line. This can be helpful as a check on individuals matched with the master chronology, i.e. since the signatures occur in most trees obviously they should occur in any new tree matched with the master. But how many signatures must an individual exhibit for a match (on this criterion) to be definite? A reflection of this problem is the lack of a computer programme to check the signatures along with the overall match. Signatures may be particularly interesting from a dendroclimatic point of view, since they represent years when a majority of trees were doing the same thing. Hence they should represent years where the signal (climate?) was in some way uniform.

## Notes

1. With incomplete samples it is often worth checking around the outer circumference to ensure that as many rings as possible have been measured.

2. Although dendrochronologists refer to measurement 'along a single radius', there is of course no rule which says that a straight line has to be followed. In practice, measurement of a sample often consists of a series of sub-radii to achieve the most regular ring pattern.

3. In particular, over-automation can be counter-productive. If the measured ring width is dumped directly on to paper tape without being seen by the operative, this excludes the running visual check on the correctness of the relative measurements from one ring to the next.

4. Since the rings are being viewed as they are measured, it is comparatively easy for a visual check to be kept on the widths being produced. If, for example, a ring is clearly wider than its predecessor but the registered widths are reversed, something has gone wrong and the rings should be remeasured.

5. It might be asked why a 'better' programme has not been written which matches on year-to-year detail, trends, etc. The answer is that if the existing programmes find matching positions, then there is only a marginal improvement in increasing the complexity of the programme. In addition, the basic assumption, which seems to be valid, that year-to-year detail does match is not necessarily true for trend matching under all circumstances. One possibility is that when all relevant reference chronologies are under control, well replicated, etc., there may be room for enhanced programmes which will compare individual ring patterns

with the reference chronologies in a number of ways — for example checking year-to-year details, trends and signatures.

6. An alternative is to plot each ring as some portion of the majority, e.g. Fn $(\frac{2}{4})$, Fn $(\frac{5}{1})$, Fn $(\frac{6}{0})$ if a suitable method can be devised.

# Chapter 4

# Modern Oaks: the Anchor in Time

It would be fair to say that in the late 1960s no one in the British Isles had a clue about the practicalities of chronology building. While information on efforts in America, Germany and Russia was available, it has to be remembered that the prevailing attitude amongst archaeologists and palaeobotanists was one of sheer scepticism. It was firmly believed that dendrochronology as a dating method would not be applicable in this maritime area. Against this background it was inevitable that some attempts to investigate the method would take the form of tests rather than direct applications of the methods which Huber had demonstrated successfully in Germany. This was particularly true of the work at Belfast, where physical distance made it unlikely that direct cross-dating could be obtained against German chronologies. On the other hand, Fletcher, who was working at Oxford, felt justified in designing his programme on art-historical dating around the assumption that trees from south-east England could be cross-dated against Germany (Fletcher *et al.,* 1974).

At Belfast it was decided to make no assumptions whatsoever but merely to investigate whether or not the basic hypothesis could be made to work reproducibly on Irish oak. Since the idea, right from the start, was to construct a chronology some six millennia in length (providing of course that the method worked at all), it was inevitable that the chronology would have to be firmly anchored in time. The obvious way to achieve such anchoring was to begin with living trees. Unfortunately, at that time it was not realised that oak trees could be successfully sampled using Swedish increment corers (these were designed for use with softwoods), so sampling was restricted to the cutting of slices using a chain saw. Since the felling of mature oaks did not seem justified purely for dendrochronology, the sampling was restricted to stumps and occasional trunks of already felled trees on various estates throughout the north of Ireland.

Some useful lessons were quickly learned. The first of these relates to the popular misconception that size or girth in an oak is proportional to age. Nothing could be further from the truth. It became obvious that the huge parkland oaks which were described as having 'three hundred rings at least' almost never exceeded a maximum of 180 years and it could only be concluded that the 300 figure was arrived at by counting right across a *diameter*. Such trees had never experienced any restriction in their growth, particularly in their canopy size. Better results, in terms of length of ring pattern, were obtained from more modest trees which had been forced by competition to stretch upwards. Frequently such trees were considerably older than their huge relatives.[1] A second lesson related to the fallibility of human memory in conjunction with an almost universal inability to judge the passage of time. Foresters and landowners would happily specify the year of felling of a stump, but these memories could only be trusted to a maximum of about two years. Beyond two years the margin of error could rapidly creep up. Sometimes three years meant five, ten meant eighteen, with the result that some samples were collected from stumps originally felled in the early 1950s.

Two examples of false pedigree should suffice. The first relates to association of particular trees with specific people. At Orange Grove, just to the south of Belfast, there is a huge Spanish chestnut to which it is reported King William III tied his horse while he rested in a nearby house. In fact it seems that originally there was a grove of these trees, but the original, to which no doubt the King did tie his horse, was blown down in the mid-eighteenth century, whereupon the reputation passed to its nearest neighbour on whose death in the nineteenth century it passed to its present incumbent. The second outlines an exercise undertaken in the United States. Dendrochronologists in the eastern states, desirous of finding long-lived trees, advertised that, as part of the bi-centenary celebrations, anyone having a tree which could be shown to have been in existence at the time of American Independence in 1776 would be issued with an appropriate plaque. Full of high hopes, they began coring some of the hundreds of trees drawn to their attention by expectant descendants of the founding fathers. It was with embarrassment that the dendrochronologists discovered that most of these reputedly old trees were seldom 100, let alone 200, years old. The outcome was that they simply handed out the plaques, said nothing, and went back to the drawing board. On this basis, what hope Robin Hood's oak?

To return to the Belfast work, the initial collection of samples yielded ring patterns which not only did not end in the same year, but for which in almost every case there was some real doubt about the actual year of felling. In particular, because it was almost never possible to get information on the season in which a tree was felled, the possibility existed that the last ring could belong to the previous year (for a spring felling) or the actual year (for a summer or autumn felling).

Although this may seem an excessively difficult way to start, it was probably one of the best things which could have happened, for the simple reason that it became necessary, right from the onset, to view the cross-dating of ring patterns objectively. In other words, we had to start by cross-dating our modern trees against one another in the same manner as we would later cross-match historic timbers. This learning exercise was to stand us in good stead later. So to review the situation with modern oaks:

(a)  they were drawn from distances up to 150 km apart;
(b)  their felling dates could not be trusted;
(c)  it was not known if missing or double rings were to be expected;
(d) it was not known if the ring patterns of Irish oaks could be cross-dated.

In practice, the plotted ring patterns were compared and similarities looked for. This was a fundamentally important stage, because if cross-matching could not be demonstrated for these trees there was indeed no hope for the method. It soon became clear that some form of confirmatory test was necessary to substantiate any matching position found. The methods used to supply back-up to human judgement, now termed visual matching, were twofold. The first centred on the whole business of replication. If two ring patterns A and B definitely matched (without doubt), then any third pattern C which matched A must also match B *at a unique position*. If it did not, then the implication had to be that C was incorrectly matched against A. As more trees were matched, each new example could be checked against A, B, C, etc. The second method of confirmation involved the use of computer programmes to check the matching positions statistically. If two ring patterns actually cross-matched, there must be something different about the matching position compared with all the other possible overlap positions: the difference being that at the matching position the ring patterns are highly correlated in a statistical sense (see Chapter 3).[2] Initially a computer programme was devised to calculate Huber's coefficient of parallel agreement (W). This was essentially identical to a programme by Eckstein and Bauch (1969). This was superseded from 1971 by the use of the Belfast CROS programme (Baillie and Pilcher, 1973).

The use of the computer programme served two purposes. If the best computer correlation coincided with an already established visual match, it gave considerable reinforcement. If, on the other hand, a better or similar correlation value was shown for a different matching position, then this could be checked visually.

As the cross-matches slowly built up, the lack of anomalies in the ring patterns suggested strongly that there were no missing or double rings. However, there were some problems, as outlined in Chapter 2, in interpreting certain configurations of spring vessels which looked like double rings. The most serious worry concerned the possibility that some rings might be missing from all trees. Although this was very unlikely (as was the possibility of all trees putting on two rings in the same year), it was something which could come through replication undetected.

As a check it was necessary to clutch at any straw which lent support. There were two factors which suggested that the ring patterns of Irish oaks represented a truly calendrical system. One was the recognition of clear signatures in the tree-rings. The other involved the coincidence of particular rings with recorded weather information.

Signatures in Irish oaks do not conform to the idea of signatures as understood by Douglass (Chapter 1). In the south-western United States a signature would be a configuration of rings, i.e. a pattern. In Irish oaks a signature refers to a single year in the sense that 1894 is a signature, i.e. every tree studied showed a narrower growth ring for 1894 than for 1893.[3] It is not necessary to recognise many such 'trend' signatures in a group of ring patterns before it becomes extremely unlikely for any individual tree to have an undetected anomaly in its ring pattern.

The second line of support concerned the coincidence of ring widths with known weather records. This is a thorny problem (see Chapter 12), as in general it is not possible to infer which factor causes any particular ring. However, in the early stages of a dendrochronological study any form of support was welcome. One of the clearest signatures in northern Irish oaks was a consistent narrow ring for the year 1816 (see Plate 1(a)). Reference to the available Irish weather records for the period showed the following entries for 1816:

> In 1816, the spring was unusually late; the summer and autumn excessively wet and cloudy . . . There were 142 wet days, principally in the summer and autumnal months. The mean temperature of the spring summer and autumn was three and a-half degrees below that of the preceeding year . . . a great want of sun and excessive rains during summer and harvest (HMSO, 1856, 175).

It later transpired that this cold, dark summer was a direct result of a massive volcanic eruption which occurred in Tambora in 1815 and gave rise to widespread cooling the following year. While it cannot be proven that the consistent narrow ring was caused specifically by these conditions, the coincidence was at least encouraging. The second coincidence was on a much more localised basis. One tree, QUB 68a from Castlecoole, Co. Fermanagh, showed clear evidence of defoliation in the year 1734. The date of this ring was arrived at on the basis of a ring count back from the fixed modern end of the chronology. Plate 2(c) shows the ring configuration for 1733 to 1736. The Irish records for 1734 contain the following entry: 'Spring very warm (and so in England) but followed by a cold and nipping May, which hurted the fruits and burned the grass' (HMSO, 1856, 122).

There is every possibility that the coincidence between the pattern in QUB 68a and the 1734 record is a causal relationship. The tree, which had been growing

**Figure 4.1: Northern Irish modern ring patterns with the Belfast index master chronology derived from 30 trees. For comparison the British Isles 18-site master is plotted below together with its five-year running mean to highlight the short- and medium-term trends.**

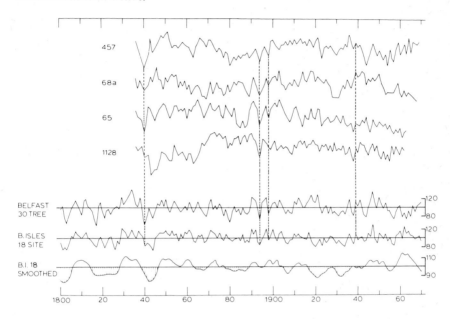

strongly in 1733, put on spring vessles as usual around May 1734 but produced no further growth that year. This was suggestive that the oaks under study represented a uniquely calendrical system.

Of the 30 oaks studied, 14 were of known cutting date. The remainder were either trunks or stumps of trees felled within the 20 years before 1970. The ring patterns of the trees with the known cutting dates showed significant cross-correlations both visually and statistically. On the basis of these agreements an initial chronology was constructed from 9 trees from two sites 150 km apart, Castlecoole and Shane's Castle. This group contained the oldest tree obtained to date QUB 528 from Shane's Castle. The resulting chronology covered 322 years, extending back in time to 1649.

The 21 other samples were each compared with this standard chronology. These comparisons were carried out for three reasons. First, to see if significant cross-correlations could be established between individual trees of known felling dates and the master, second to establish the dates of the outer years of samples which were otherwise undated, and, third, to observe the range of 't' values established as a guide to later work. The apparent age of some of the samples is accounted for by the fact that some trees exhibited bands of narrow rings or areas of physical deterioration which precluded the measuring of their total ring patterns.

**Table 4.1: Correlations between Modern Samples and a Preliminary Master Chronology**

| | | QUB No. | Length (years) | 't' | Outer Ring |
|---|---|---|---|---|---|
| Pomeroy | I | 65 | 170 | 8.12 | 1963 |
| Springhill | I | 1059 | 103 | 2.82 | 1836 |
| | II | 1060 | 151 | 5.17 | 1900 |
| Moneymore | I | 532 | 83 | 3.79 | 1969 |
| Draperstown | I | 558 | 140 | 4.85 | 1968 |
| Antrim | I | 457 | 135 | 6.74 | 1969 |
| Barnetts Pk. | I | 466 | 178 | 6.20 | 1970 |
| | II | 467 | 111 | 7.50 | 1970 |
| | III | 462 | 108 | 7.41 | 1954 |
| | IV | 456 | 150 | 3.50 | 1922 |
| Dixon Pk. | I | 531 | 193 | 5.57 | 1969 |
| Cultra | I | 41 | 144 | 4.47 | 1967 |
| | II | 44 | 120 | 4.21 | 1967 |
| | III | 45 | 120 | 3.89 | 1967 |
| Saintfield | I | 1129 | 178 | 3.54 | 1957 |
| | II | 1128 | 140 | 3.78 | 1962 |
| Castleward | I | 66b | 143 | 4.82 | 1966 |
| | II | 66a | 143 | 4.46 | 1966 |
| Tollymore | I | 67a | 208 | 5.18 | 1966 |
| | II | 67b | 244 | 4.78 | 1966 |
| | III | 67c | 132 | 2.34 | 1966 |

With the exception of QUB 67c and 1059, the 't' values obtained were all significant. This shows that a chronology produced from as few as 9 trees contains a reliable record of the relative quality of each growing season within the area of study. Of the two examples which produced low significance levels, QUB 67c produced t values of 5.17 and 6.44, with its outer year at 1966, when cross-correlated with QUB 67a and 67b from the same site. Figure 4.1 shows the ring patterns for four modern trees from widely differing locations within the area of study. A number of signature years are marked at 1840, 1894, 1898 and 1939. Over the period from 1800 to 1970 a total of 20 years exhibit 90 per cent agreement in trend over the 30 component trees.

It was clear that the modern oaks in the north of Ireland were artificially planted. Most came from estates set up in the eighteenth and early nineteenth centuries, and their ages reflected the planting of trees by landlords in those centuries. Only one source, Shane's Castle, Co. Antrim, yielded oaks which had started growth in the seventeenth century — and then only two examples. In all subsequent searching no older living oaks have been found in Ireland. There do not appear to be any substantial remains of the primeval forests of Ireland extant at the present day. It will be suggested later that the so-called primeval woods which existed in the seventeenth century (McCracken, 1971) may well have been the result of earlier abandonment of land to forest rather than virgin forest (see Chapter 7).

The construction of the chronology for the period 1649 to 1970 and the establishment of statistical values for cross-correlations between ring patterns gave some feel for the method. The high, usually uniquely high, correlation values lent hope that extension of the modern chronology would be possible.

## Irish Site Chronologies

One serious shortcoming of the early work described above was the failure to test the degree of correlation between trees or chronologies from different areas within Ireland. Instead of testing to see if the country was made up of multiple tree-ring areas[4] or a single tree-ring area, an assumption was made that, as a small geographical unit, the north of Ireland should form an area within which cross-agreement between ring patterns could be expected. Clearly this assumption was borne out by the modern results above, but it meant that no information existed on whether trees from the southern half of Ireland would match with the Belfast chronology. The situation within the British Isles as a whole was even less well understood.

No serious move was made to rectify this situation until 1978, when as a first move in the construction of a grid of modern chronologies it was decided to build a series of highly localised oak chronologies from diverse parts of Ireland. This work was carried out by Jon Pilcher and the author as part of a contribution towards dendroclimatological research (see Chapter 12). By 1978 the sampling of oaks had been simplified by the availability of increment corers. Figure 4.2 shows the locations of the sites chosen (essentially at random). At each site at least 10 trees were sampled and a site chronology constructed. This rapid step from individual trees to site chronologies was aided by the exact temporal control of simultaneous sampling, each tree's growth ending in the same year.

Having established seven site chronologies, these could be compared in a number of ways. Table 4.2 shows the cross-correlation *t* values between each of the site chronologies for the common period 1850 to 1969.

Figure 4.2: Locations of the 18 modern chronologies cited in the text. The shaded area is the source of the Northern Irish trees used in the Belfast master and represents one-twentieth of the overall 18-site area. The dotted line separates areas with more than 30 inches rainfall (west).

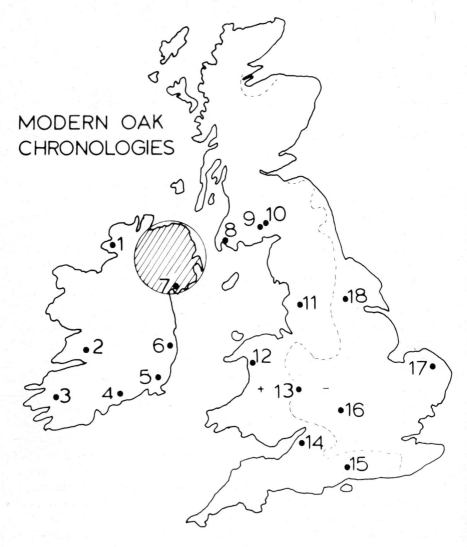

MODERN OAK CHRONOLOGIES

**Table 4.2: *t* between Site Chronologies and Distance for the Common Period 1850 to 1969**

| Ardara | Ar | R | GD | E | C | K | LD | Distance (Arbitrary units) | | | | | |
|---|---|---|---|---|---|---|---|---|---|---|---|---|---|
| Rostrevor | 2.5 | | | | | | | 11.0 | | | | | |
| Glen of Downs | 2.4 | 3.2 | | | | | | 15.5 | 7.0 | | | | |
| Enniscorthy | 1.3 | 5.4 | 1.3 | | | | | 19.0 | 12.0 | 5.5 | | | |
| Cappoquin | 3.2 | 1.4 | 4.2 | 4.4 | | | | 20.0 | 16.5 | 11.0 | 6.5 | | |
| Killarney | 4.5 | 5.5 | 3.1 | 6.9 | 3.8 | | | 21.0 | 21.5 | 17.5 | 14.0 | 7.5 | |
| Lough Doon | 7.3 | 5.4 | 2.7 | 4.8 | 3.0 | 7.1 | | 14.5 | 15.0 | 12.0 | 10.0 | 6.5 | 7.0 |
| Belfast | 10.3 | 6.0 | 2.9 | 4.8 | 2.9 | 4.6 | 7.4 | | | | | | |

If the correlation values are plotted against distance, it is clear that no real relationship exists (see Figure 4.3). On the basis of this it can be suggested that the differences between the chronologies are *site* differences. That is, all the sites are responding to the same basic signal which is responsible for the cross-agreement between their ring patterns. This signal is not affected by geographical location but merely by on-site factors. This finding could have important implications for assigning the whole of Ireland to a single tree-ring area.

The bottom line of Table 4.2 shows the degree of correlation between each of the site chronologies and the original Belfast generalised chronology whose construction was described earlier in this chapter. If a reasonable geographical centre point for this general chronology is Lough Neagh, then again there is no evidence for correlation values falling off with distance. With the exception of the Glen of the Downs and Cappoquin chronologies, each of the other matches is highly significant or better, and each of the chronologies would have been datable against the Belfast chronology even if their sampling dates had not been known. To follow up this line of enquiry, the information contained in the seven site chronologies was combined into an overall Irish master chronology for the period 1850 to 1969. Each of the site chronologies was then compared with the Irish master in the same manner as above and for the same common period. The resultant *t* values are listed in Table 4.3 together with the comparison between the Belfast and Irish chronologies. There are several observations which can be made on these results. Obviously each of the chronologies shows extremely significant agreement with the Irish master chronology; however, it is clear that Glen of the Downs and Cappoquin show the least good agreement. Since they are constituents of the Irish chronology and carry equal weight in its construction, it can be deduced that these two sites are recording the overall climatic signal least well.

**Figure 4.3:** *t* values (common period) plotted against distance for the 7 Irish modern chronologies, showing no noticeable correlation.

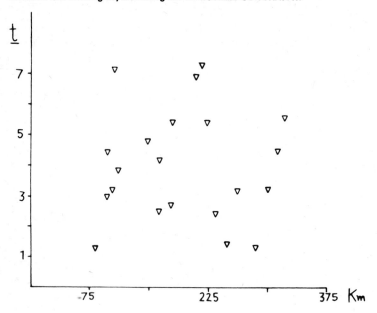

**Table 4.3: Each Site Compared with the Irish Master Chronology**

| Site | *t* (1850-1969) | *t* (1850-1909) | *t* (1910-69) |
|------|-----------------|-----------------|---------------|
| Belfast | 9.2 | 6.4 | 6.5 |
| Ardara | 10.5 | 9.4 | 5.6 |
| Rostrevor | 12.4 | 7.2 | 10.4 |
| Glen of the Downs | 6.8 | 4.3 | 5.1 |
| Enniscorthy | 8.2 | 4.5 | 7.3 |
| Cappoquin | 7.8 | 5.1 | 5.6 |
| Killarney | 12.3 | 10.6 | 6.8 |
| Lough Doon | 11.6 | 8.4 | 7.8 |

Whether this is due to purely geographical factors or some more localised site factors requires further study. This finding does help to explain why these two sites in particular showed the lowest significance levels when compared with the Belfast chronology above.

The high level of agreement between the Irish and Belfast chronologies ($t = 9.2$) suggests that we are dealing with a generalised signal controlling tree growth in Ireland. It could be suggested that site chronologies tend to enhance the 'differences' due to localised site factors while the overall average signal remains similar. This seems to be the only way to explain the similarities between the Belfast chronology, with its northern sources, and the Irish master, which should have a bias towards the more numerous southern sites. One implication of this might be that, in the future, when sufficient information is available from a number of areas, it may be possible to construct a single 'average' chronology useful for dating within the whole of Ireland.

These observations tend to suggest a picture which may be slightly over-optimistic. The result of building these modern chronologies has been a slightly better understanding of the level of agreement existing between trees from different areas *in recent times*. However, because there is agreement between, for example, the extreme south-west and north-east of Ireland in one (the modern) period does not mean that such agreement *must* be expected in other periods. To illustrate this, the cross-correlation of each site chronology with the overall Irish master was also carried out in two sections:

(a) for the period 1850 to 1909.
(b) for the period 1910 to 1969.

Since the site chronologies contained only the information from identically the same trees in each period, it is interesting to note the differences between the two sets of results in Table 4.3. Coming forward in time, two of the western sites, Ardara and Killarney, show a reduced agreement with the overall master, while two of the eastern sites, Rostrevor and Enniscorthy, show an enhanced agreement, the others remaining broadly constant. The suggestion has to be, since the trees themselves can have changed very little, that some other, presumably climatic, factor has *changed with time*. This line of thinking introduces an imponderable into our considerations, viz. even if diverse timbers can be dated against the Belfast chronology for some periods, there is no guarantee that the same will apply for all periods. Having said this, it is clear that a high level of agreement can obtain between master chronologies from diverse areas within Ireland.

**Modern Oak Chronologies in Britain**

Most modern oaks in England and Wales are the result of planting in the eighteenth and nineteenth centuries. This is a similar situation to that noted in Ireland and it is clearly shown in the maximum extent of available English modern chronologies listed in Table 4.4.

**Table 4.4: English Chronologies and their Earliest Extent Using Living Trees**

| Chronology | Extent | Source |
|---|---|---|
| Winchester | 1635 | Barefoot, 1975 |
| Cumberland — Hereford | 1729 | Siebenlist-Kerner, 1978 |
| Yorkshire (?) Cumberland | 1710 | Morgan (personal communication) |
| Maentwrog (Wales) | 1710 | Leggett *et al.*, 1978 |
| Beechen | 1823 | Fletcher, 1974 |
| East Midlands | 1459 | Laxton *et al.*, 1979 |

Obviously there are some very long-lived oaks in Britain, but they form a very small proportion of the overall oak population. For example, the East Midland start date 1459 needs to be qualified in this context as only 3 trees go back before 1700. In 1976-7, in the course of constructing an oak chronology for southern Scotland, modern oaks were sampled at two ancient woodlands; Lockwood, Dumfriesshire and Cadzow, near Hamilton. The former yielded ring patterns back to 1571, the latter to 1444. It must be stressed that these woodlands are exceptional. Attention was drawn to them and they were sampled solely because of their great age. Interestingly, at both sites the ages of the trees, while considerable, fell well short of local expectations. Cadzow in particular was reputed to be of the order of 800 years old but, while the woodland may be, it is doubtful if any of the trees presently on the site are anything like that age. Of the ten stumps sampled at Cadzow only one ran back to 1444. Five of the remainder had begun life within a decade of 1500.[5] This clustering, 50 per cent of a random sample being effectively identical in age, is suggestive of a major regeneration phase at that time. It is possible that around 1500 the area was enclosed as a cattle park and subsequent regeneration was curtailed. This would have the effect of yielding a significant proportion of trees from the period immediately before the enclosure[6] (Baillie, 1977a, 33).

*Correlations between British Isles Chronologies*

In 1977 there were only five modern British Isles chronologies covering the common period 1710 to 1970. These were the chronologies for Belfast, southern Scotland, Yorkshire, Maentwrog and Winchester. At that time each was compared with all the others and with a shorter chronology of Beechen (1823-1971) (Baillie,

1978a, 36). Table 4.5 shows the correlation values between these chronologies. With the exception of Maentwrog *v.* Yorkshire all the others show strong positive correlation. This was encouraging, as it lent hope that material from different areas could be cross-dated. In particular it indicated that tree-ring patterns from diverse areas within the British Isles were not mutually exclusive. This was important because Fletcher, on the basis of his art-historical chronologies, was postulating mutually exclusive tree-ring areas called Type A and Type H (Fletcher, 1977, 1978a).

In 1978 and 1979 a further 7 modern chronologies were constructed at Belfast using English and Scottish oak cores. These were again aimed at dendroclimatology. By 1979 it was possible to look at a total of 18 modern chronologies for the common period 1850 to 1969. Their distribution is shown in Figure 4.2. On the basis of what had been done in Ireland, i.e. the construction of an all-Ireland chronology, it was decided to construct one British Isles chronology using all 18 sites. These stretched from Killarney to Norwich and from Dumfries to Winchester. Each chronology was given equal weighting by the use of indices rather than mean ring widths. The total number of trees involved was of the order of 250. The question has to be asked: 'What was expected from an exercise of this kind?'

Clearly no one chronology could bias the overall master. If all the chronologies were random, i.e. if there was no common signal, the average would be expected to come out close to a straight line. In practice the 18-site master looked like a perfectly normal tree-ring pattern with clear year-to-year variations and short- to medium-term trends. The chronologies are not cancelling one another out! In particular the occurrence of trends is important. In any index chronology only trends consistently present in the individual trees will come through to the final master. Similarly, with the 18-site master only consistent trends in the site masters would come through to the overall British Isles master. For clarity a five-year moving average is drawn below the master in Figure 4.1. Since the trends must be real the general tendency to narrow rings in the 1920s and 1930s must be real: similarly the tendency to wider rings in the 1940s and 1960s. It appears that this British Isles master has both real detail and features. An obvious test is to compare each of the constituent site chronologies with the overall master. Simplistically this is equivalent to asking, 'Which site chronology is most like the overall chronology?' Clearly the answer could not be predicted in advance.

The correlations between the site masters and the overall master were calculated in two ways. First, the CROS programme was used. This programme as outlined in Chapter 3, contains a powerful filter which excludes all but the highest-frequency information, i.e. it correlates the year-to-year detail and ignores trends in the data. The second set of correlations used CROS with the filter removed; in this case the correlation was performed on data with short and medium trends included.[7]

**Table 4.5: Correlation *t* Values between 'Local' Chronologies for the Period AD 1710 to 1970, Maentwrog (Leggett *et al.*, 1978), Winchester (Barefoot, 1975), Beechen (Fletcher, 1976), and Yorkshire (Morgan, personal communication).**

|  | Belfast | Scotland | Maentwrog | Yorkshire | Beechen |
|---|---|---|---|---|---|
| Scotland | 10.1 | | | | |
| Maentwrog | 5.4 | 3.4 | | | |
| Yorkshire | 7.1 | 5.9 | 1.2 | | |
| Winchester | 4.3 | 3.6 | 5.1 | 4.6 | |
| Beechen 1823-1971 | 3.1 | 2.5 | 3.5 | 3.9 | 4.9 |

The *t* values produced vary slightly but the overall picture remains remarkably constant. Each site chronology shows strong agreement with the overall master. It would seem, therefore, that the master must be a reasonable reflection of some general signal which the trees are receiving. When the correlation figures are plotted on a map of the British Isles an interesting picture emerges (see Figure 4.4). The distribution of high and low values is not random. The highest values in both cases are associated with the western sites. Therefore, whatever the signal is, it is the western sites which are receiving it most strongly. The difference between the filtered and unfiltered correlation is worth noting. It appears that both the year-to-year detail and the trends related to the overall signal are received best at the westerly sites. When trends are taken into account the most easterly site at Norwich shows markedly less agreement with the overall master. This may lend support to Fletcher's suggestion that the North Sea influences sites in the east of Britain (Fletcher, 1978a, 153). However, there can be no doubt that all the sites examined show a strong response to the overall signal, particularly in year-to-year detail.

This would seem to be a fundamentally important observation. If enough sites are lumped together it is possible to produce a tree-ring chronology which is valid, for example for dating purposes, over a wide area. This might imply that in the future it will be possible to produce generalised chronologies which will date most site masters and this in turn could influence any approach to the creation of a grid of tree-ring chronologies. The reader may ask whether the inclusion of each site in the overall master could have given rise to the high correlation values? There are two replies to this, one intuitive, i.e. how could one chronology out of eighteen possibly bias the overall master? The second is more empirical There are other modern chronologies available, in particular the modern Belfast chronology

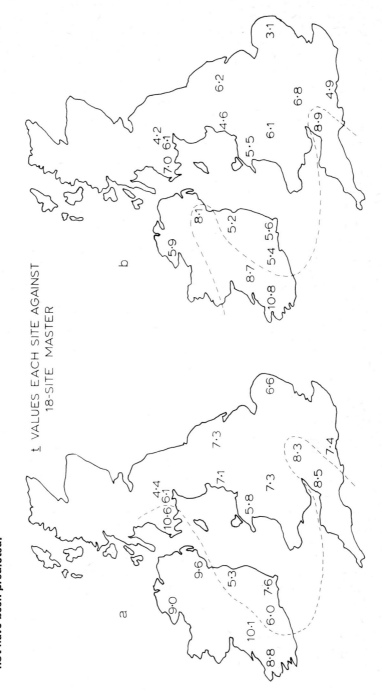

Figure 4.4: (a) *t* values between each site chronology and the overall 18-site master using the standard CROS programme. (b) *t* values when trends are taken into account (using CROS without its high-pass filter). In each case the highest values occur in the west, indicating a better response to the common signal in those areas, a result which could not have been predicted.

**Figure 4.5: An example of the standard CROS output — in this case running the 1850 to 1979 section of the 18-site master against the published Belfast chronology (Baillie, 1973b). Note the lack of values greater than $t = 3.5$ except for the matching position.**

COMPARISON OF TREE BELFAST BELIM 15 AND ENG/IRL/MASTER CHRON EIGHTEEN

| DATE OF OUTER YEAR | 'T' VALUES FOR COMPARISONS | | | | | | | | | |
|---|---|---|---|---|---|---|---|---|---|---|
| 1565 | 0.00 | 0.55 | 0.00 | 0.00 | 1.61 | 0.00 | 0.00 | 1.26 | 0.00 | 0.00 |
| 1575 | 2.36 | 0.24 | 0.00 | 0.00 | 1.55 | 0.00 | 0.00 | 0.14 | 0.00 | 0.00 |
| 1585 | 0.13 | 0.78 | 0.00 | 0.26 | 0.00 | 1.04 | 1.34 | 0.00 | 0.57 | 0.08 |
| 1595 | 0.28 | 0.00 | 0.00 | 1.09 | 0.03 | 0.00 | 1.57 | 0.00 | 0.00 | 0.23 |
| 1605 | 0.00 | 1.54 | 0.00 | 0.00 | 1.77 | 0.00 | 0.00 | 1.98 | 1.23 | 0.00 |
| 1615 | 0.00 | 0.00 | 0.87 | 0.30 | 0.00 | 0.00 | 1.79 | 0.00 | 0.88 | 1.74 |
| 1625 | 0.00 | 0.00 | 1.46 | 0.79 | 0.00 | 0.00 | 0.86 | 0.57 | 0.00 | 0.75 |
| 1635 | 0.00 | 0.00 | 0.76 | 0.20 | 0.85 | 0.00 | 0.00 | 2.00 | 0.29 | 0.00 |
| 1645 | 0.00 | 1.43 | 0.33 | 0.00 | 0.72 | 0.83 | 0.00 | 0.00 | 1.24 | 0.56 |
| 1655 | 0.00 | 0.57 | 0.00 | 0.80 | 0.74 | 0.00 | 0.00 | 0.10 | 1.10 | 0.05 |
| 1665 | 0.00 | 0.00 | 0.00 | 0.11 | 1.00 | 0.00 | 0.00 | 0.00 | 1.75 | 0.60 |
| 1675 | 0.18 | 0.00 | 0.00 | 0.95 | 0.00 | 0.00 | 0.63 | 0.02 | 0.00 | 1.96 |
| 1685 | 0.00 | 0.00 | 0.64 | 1.02 | 0.12 | 0.00 | 0.00 | 0.82 | 1.05 | 0.00 |
| 1695 | 0.00 | 0.00 | 0.52 | 1.45 | 1.37 | 0.00 | 0.00 | 2.05 | 0.00 | 0.00 |
| 1705 | 0.67 | 0.91 | 0.00 | 0.74 | 0.00 | 0.00 | 1.61 | 0.46 | 0.00 | 0.00 |
| 1715 | 0.00 | 1.73 | 1.58 | 0.00 | 0.00 | 0.09 | 0.14 | 0.00 | 0.43 | 0.26 |
| 1725 | 0.53 | 0.00 | 0.00 | 0.44 | 0.72 | 0.00 | 0.00 | 0.00 | 0.00 | 2.52 |
| 1735 | 0.00 | 0.00 | 2.20 | 0.75 | 0.00 | 0.00 | 0.48 | 0.59 | 0.00 | 0.57 |
| 1745 | 0.00 | 0.24 | 0.06 | 0.38 | 1.10 | 0.00 | 0.20 | 1.30 | 0.00 | 0.00 |
| 1755 | 1.43 | 0.61 | 0.00 | 0.00 | 0.38 | 0.00 | 0.00 | 0.00 | 1.36 | 0.31 |
| 1765 | 0.00 | 0.46 | 2.15 | 0.09 | 0.00 | 0.00 | 0.00 | 0.68 | 1.33 | 0.00 |
| 1775 | 0.07 | 0.36 | 0.00 | 1.46 | 0.00 | 0.00 | 2.24 | 0.77 | 0.00 | 0.00 |
| 1785 | 0.00 | 2.48 | 0.90 | 0.00 | 0.00 | 0.00 | 0.36 | 0.73 | 0.00 | 0.00 |
| 1795 | 1.66 | 0.00 | 0.00 | 1.20 | 0.40 | 0.70 | 0.00 | 0.00 | 0.16 | 0.00 |
| 1805 | 0.49 | 2.69 | 0.00 | 0.00 | 0.00 | 0.00 | 0.00 | 0.86 | 1.61 | 0.00 |
| 1815 | 0.00 | 0.08 | 1.55 | 1.86 | 0.00 | 0.00 | 0.76 | 0.00 | 0.00 | 0.63 |
| 1825 | 0.00 | 0.11 | 1.54 | 0.00 | 0.00 | 0.08 | 2.71 | 0.73 | 0.00 | 0.00 |
| 1835 | 0.48 | 0.00 | 0.00 | 1.43 | 1.31 | 0.00 | 0.00 | 0.57 | 1.37 | 0.00 |
| 1845 | 0.00 | 0.00 | 2.05 | 0.00 | 0.00 | 0.49 | 1.87 | 1.07 | 0.00 | 0.00 |
| 1855 | 1.56 | 1.41 | 0.00 | 0.26 | 0.00 | 0.00 | 0.20 | 0.82 | 2.25 | 0.00 |
| 1865 | 0.00 | 0.00 | 2.10 | 0.00 | 0.00 | 0.86 | 0.00 | 0.00 | 0.22 | 0.36 |
| 1875 | 0.03 | 0.76 | 1.43 | 0.00 | 0.00 | 1.15 | 2.59 | 0.59 | 0.00 | 0.00 |
| 1885 | 0.50 | 0.52 | 0.85 | 0.33 | 0.00 | 1.06 | 0.00 | 0.00 | 1.04 | 1.49 |
| 1895 | 0.00 | 0.00 | 0.10 | 1.16 | 0.00 | 0.00 | 0.72 | 1.27 | 0.00 | 0.70 |
| 1905 | 1.30 | 0.00 | 1.05 | 0.79 | 0.00 | 0.00 | 0.21 | 0.38 | 0.35 | 0.00 |
| 1915 | 0.00 | 1.07 | 0.00 | 1.22 | 0.00 | 0.00 | 0.46 | 0.33 | 0.00 | 0.00 |
| 1925 | 0.89 | 1.02 | 0.00 | 0.00 | 1.06 | 0.83 | 0.00 | 0.67 | 1.24 | 0.00 |
| 1935 | 0.00 | 0.32 | 0.45 | 0.00 | 0.34 | 0.54 | 0.00 | 0.00 | 0.54 | 0.16 |
| 1945 | 0.00 | 0.48 | 1.05 | 0.00 | 0.00 | 1.40 | 0.00 | 0.00 | 1.03 | 1.19 |
| 1955 | 0.00 | 0.00 | 0.00 | 0.49 | 0.00 | 0.05 | 1.16 | 0.09 | 0.00 | 0.31 |
| 1965 | 1.57 | 0.00 | 0.00 | 1.53 | 0.00 | 0.00 | 0.86 | 0.00 | 0.00 | 1.16 |
| 1975 | 1.82 | 0.00 | 0.00 | 0.00 | 11.26 | 0.00 | 0.00 | 0.00 | 1.96 | 0.16 |
| 1985 | 0.24 | 0.00 | 0.05 | 0.00 | 0.00 | 1.70 | 0.00 | 0.00 | 0.57 | 0.00 |
| 1995 | 0.00 | 1.16 | 0.55 | 0.00 | 0.00 | 1.40 | 0.00 | 0.00 | 1.47 | 3.08 |
| 2005 | 0.00 | 0.00 | 0.00 | 1.53 | 0.77 | 0.14 | 0.61 | 0.00 | 0.00 | 1.39 |
| 2015 | 0.93 | 0.00 | 0.55 | 0.02 | 0.00 | 0.00 | 1.31 | 0.93 | 0.00 | 0.00 |
| 2025 | 1.93 | 0.00 | 0.00 | 1.24 | 3.44 | 0.41 | 0.00 | 0.00 | 0.00 | 1.33 |
| 2035 | 0.65 | 0.00 | 0.00 | 0.00 | 1.11 | 0.24 | 0.00 | 0.67 | 1.07 | 0.00 |
| 2045 | 0.00 | 0.00 | 0.43 | 1.49 | 0.56 | 0.00 | 0.00 | 0.00 | 2.11 | 2.20 |
| 2055 | 0.45 | 0.00 | 0.00 | 0.00 | 1.97 | 0.59 | 0.17 | 0.00 | 0.00 | 0.00 |
| 2065 | 0.00 | 0.06 | 1.92 | 0.49 | 0.00 | 0.00 | 0.00 | 2.23 | 0.70 | 0.00 |
| 2075 | 0.00 | 0.02 | 0.35 | 1.03 | 0.25 | 0.33 | 0.00 | 0.00 | 0.00 | 1.39 |

'T' =*11.26 ,DATE OF OUTER YEAR = 1979

described above. It is not included in the overall master, yet it gives a correlation value of $t = 11.3$ against the overall master for the common period 1850 to 1979. The remarkable similarity between these chronologies is shown in Figure 4.1. The similarity would argue, since the Belfast master is made up from trees from widespread sources within the north of Ireland, that in each case the combination of trees from different sites rapidly produces the basic signal. Both of these chronologies were made by exactly the same procedures and are presented in identical format, yet they were constructed using different samples from areas twenty times different in size. The relevant CROS output is shown in Figure 4.5.

The results discussed above tend to paint a rather rosy picture, in the sense that they suggest a relatively simple tree-ring situation. However, a cautionary word is necessary. It is possible that the whole pattern can change with time. If we go back to the Scottish modern chronology, composed of Lockwood and Hamilton trees, and compare it with the Belfast chronology for the eighteenth and nineteenth centuries, the correlation results are $t=4.3$ and $t=11.4$ respectively. It is possible that the period back to 1800 has been a period when prevailing conditions favoured similarities between ring patterns. It must also be possible that under different conditions dissimilarities might be favoured. Only empirical observation will give the answer.

## Notes

1. This led the author to coin the somewhat exaggerated truism that with regard to oaks 'the bigger they are, the younger they are'.

2. If this were not true then it would be impossible to pick out the matching position in any repeatable fashion and the whole basis of the method would collapse.

3. Huber designated such majority years as signatures if 75 per cent or more of all trees showed the same trend.

4. For all anyone knew, agreement between trees might have been an extremely limited phenomenon in a geographical sense.

5. A report on the Cadzow oaks by tree surgeons in the 1970s estimated the trees to be around 250 years old on the basis of size. The ages discovered during tree-ring analysis suggest that most of the trees are almost twice this age.

6. Up until 1978 the last remnants of a herd of white 'Celtic' cattle were kept in the Cadzow woodland. While no young oaks were evident within the enclosure, abundant saplings could be found just a few metres beyond the fence. This would seem to confirm that the cattle were responsible for suppressing regeneration.

7. Since each index chronology varies about a common axis, using a straight filterless correlation compares the trends as well as the year-to-year detail.

# Chapter 5

# Extension of the Belfast Modern Chronology

It is important to understand the context of this chronology extension. At Belfast in the early 1970s it was not possible to cross-date chronologies against any outside reference chronology. Thus the logical extension of the modern chronology covering 1970 to 1649 had to be totally independent. To understand fully the problems involved it is necessary to set the scene historically. In particular the history of buildings is not the same in this area as in the more settled context of England.

## Historical Factors Governing Timber Availability in the North of Ireland

Historical events in the north of Ireland have had a decisive effect on the survival of buildings and hence on the availability of timbers. It would be useful to examine the significant factors in the establishment of the Northern Irish community and, in particular, the dates associated with wholesale destruction of property in the seventeenth century.

When the Normans invaded Ireland in the late twelfth century they established a chain of outposts around the country, first in motte-and-bailey castles, still significant features of the Irish countryside, and later in strong, stone-built castles of which Carrickfergus Castle in Co. Antrim is one of the best preserved. These castles were however little more than beach-heads for incursions into the surrounding areas. The north of Ireland was not successfully conquered by the English forces until the reign of Elizabeth I. The victory in this Tudor campaign did not come until 1603, and the final acknowledgement of this defeat is accepted as the Flight of the Earls in 1607, when ninety of the leading men of Ulster went into exile on the Continent (Clarke, 1967).

The scheme devised at this time, to ensure the continuing obedience of this northern part of Ireland, was to plant colonists from England and Scotland on to the land and dispossess the native Irish. As a lever to encourage people to move to Ireland James I paid his creditors with grants of Irish land. The result was the Plantation of Ulster with a large body of settlers, at the expense of the native

Irish. The settlers brought with them the techniques of building common at the time in their native counties. So, for the first time as far as is known, substantial timber-framed houses were built in the north of Ireland. Previous to the Plantation, apart from the few tower houses and castles belonging to Irish noblemen, the habit of the Irish builders had been the construction of non-permanent buildings in consequence of their partially nomadic existence (Paterson, 1960).

In the first major attempt by the Irish to drive out the settlers, the Rebellion of 1641, their main action was the destruction of houses and castles. In the first sweep of the Rebellion, the whole of the lands west of the River Bann, i.e. the counties of Londonderry, Tyrone, Fermanagh and most of Armagh fell into Irish hands. Almost all of the timber-framed buildings in these areas were destroyed. The remaining two counties did not get off lightly. In Co. Down, Dromore and Lisburn were burnt.

The initial success of the Rebellion was soon reversed, although it was not until the arrival of Cromwell with his experienced Roundhead army that the rising was completely crushed. As a result of the 1641 Rebellion more land was confiscated and the areas where the settlers had been driven out were re-planted.

The Rebellion had lasted from 1641 to 1653, but during the whole period of the Commonwealth, up to the restoration of Charles II in 1660, Ireland suffered a period of extreme depression. Several reasons combined to cause this situation, among them the devastation left by the Rebellion, external economic factors and a debased coinage (Camblin, 1951, 48). Only after the return of the monarchy in 1660 was there a full revival of confidence and a new phase of building, both secular and religious. The land was again in the hands of planters and the forests were exploited not only for the rebuilding of houses and churches but also to supply the iron industry and the trades of coopering and tanning (McCracken, 1971).

The succession of James II in 1685 marked the beginning of a second period of unrest in Ireland. This closely paralleled the situation in England, where the great landowners were seeking to limit the authority of the Crown. Lees-Milne (1970) states that the unsettled state of the country between 1685 and 1688 can be gauged by the pause in country-house building.

In 1688 William III arrived at the request of the English people, to replace James II. James was forced to flee to Ireland, where he was supported by the Irish who saw in his cause the opportunity of re-establishing themselves on the lands which they had fought to regain in 1641. For a time James's army had control of most of the north of Ireland. Camblin (1951, 49) states that

the whole of the countryside in the western part of Ulster was devastated at this time, the settlers having fled to places of safety . . . large areas were left without inhabitants and such was the state of desolation that the Jacobite army could scarcely obtain food or shelter.

James's army was gradually driven southwards, by the forces of William, until its final defeat in 1690. The result was the exile of James to France and the establishment of a settled monarchy which favoured the English settlers in Ireland.

The events of the seventeenth century can be seen to have had a decisive effect on the availability of historic timbers in the north of Ireland. Virtually no buildings exist from the period before 1641. Certainly no medieval timber houses remain and none of the introduced English frame-houses of the seventeenth century have been discovered (Evans, 1957). Little building can have taken place between 1641 and the mid-1650s, though undoubtedly the few years following the Restoration must have seen considerable activity. Many of the buildings erected after 1660 were destroyed during the period 1688-90 and, in consequence, after 1690 there was another concentrated building phase. In the area of study, most of the surviving buildings containing native oak timbers should belong to this late seventeenth century period.

The previous statement is based on the fact that while few early oak-built houses survived the seventeenth century, very few were built after its close. The main reason for this would seem to be the shortage of oak timber due to the disappearance of the oak woods. This in turn was the result of the industrial exploitation mentioned above, which was exaggerated between 1685 and 1688 by wholesale cutting in fear of land confiscation. That the situation was becoming increasingly serious through the seventeenth century is shown in various references. Boate, writing in 1654, commented on the disappearance of large tracts of forest in Co. Down and stated that the north part of Tyrone was the only wooded part of the country (Hutchinson, 1951). McCracken (1947) states specifically that by the end of the seventeenth century the greater part of Ulster's woods were gone. A reflection of the increasing scarcity of oak timber for building is seen in the growing use of imported and fossil timber after 1700. Camblin (1951, 59) states that the destruction of the woods during the whole of the seventeenth century was so great that in 1726 a considerable amount of timber was being imported for building and in 1729 'tis a great misfortune that we are under the necessity of sending to distant countries for building our houses'.

All of the known eighteenth-century buildings investigated in the north of Ireland were constructed using imported pine. This applied equally to domestic and industrial buildings. Poorer houses frequently made use of fossil timbers from bogs. Mason's Parochial Survey of Ireland in 1816 states that 'the cottages are . . . roofed with foreign timbers as bog oak which formerly supplied them is nearly exhausted' (Paterson, 1960).

So in the area under study there were almost no extant buildings containing oak timbers felled before 1600. The two serious uprisings of 1641 and 1688 substantially reduced the number of oak-framed buildings which survived the seventeenth century. In addition, the seventeenth century saw an almost complete destruction of the natural oak forests, with the result that from a date between 1720 and 1750 virtually no forest oak remained for use in building construction.[1] As a result, from the early eighteenth century use had to be made of imported or fossil timbers for house construction.

## Extension of the Modern Chronology

Obviously the extension of a modern chronology is dependent on the existence of timbers of earlier date from the same area. The criterion for usefulness is that some part of the age range of the earlier timbers must overlap that of the oldest modern trees. The earliest extension of any available modern tree was 1649. For a reliable cross-correlation to be established between two ring patterns an overlap of more than 70 years was desirable. On this basis it was necessary to find timbers felled after 1725 for use in chronology extension. While at first sight this does not seem a particularly stringent criterion, the unusual history of the north of Ireland made it difficult to fulfil.

When the tree-ring project began in 1968, virtually no background information existed about available timbers from historic buildings. Further, in published articles concerning house types, dates were established on the basis of style, building technique, written or verbal evidence. This type of information, while generally correct in assigning a building to a particular period, seldom claimed a dating accuracy better than a half-century. In addition, because only derelict houses had been investigated for publication by vernacular architects, most of the published vernacular buildings had already disappeared by 1968.

In the search for eighteenth-century oak timbers an empirical rule soon developed. If a building bore a definite construction date, it was found to be of pine construction, i.e. imported timber had been used for its wooden parts. This is partially explained by the fact that only the more important houses have definite building dates. This is certainly true of all of the large Georgian houses, for example Castlecoole (1788), Saintfield (1750), Castleward (1763), Baronscourt (1793, where building was delayed specifically by the late arrival of a shipment of timber), Downhill (1780), and Government House at Hillsborough (1795). Imported pine was also used in less impressive houses as well as in all recorded mills and industrial buildings throughout the eighteenth century. The corollary of this was that when a building in the course of repair or demolition yielded oak timbers, in general, little specific dating evidence existed.

The inherent problem in working with timber of unknown age, in this situation, is that it is not known whether any part of its ring pattern and that of the modern chronology grew over the same period. Further, at best, any overlap was likely to be short. Thus, the approach to chronology extension was dictated entirely by the available material.

Basically, the approach had to be similar to that used by Douglass and his co-workers in America prior to 1929. Using timbers from sites of unknown antiquity, he extended a floating sequence forward in time towards the present to link the prehistoric sequence with a living-tree chronology (see Chapter 1). The analogy between that work and the Belfast project was very close. The majority of oak samples available in the north of Ireland came from abandoned or ruinous buildings with little clue as to their age. Cross-correlations with long overlaps are more likely to be obtained between samples of roughly the same period than between the extremities of samples of radically differing periods. Hence the approach to chronology extension at Belfast was to cross-correlate ring patterns from timbers of unknown age and hope that eventually the resultant floating sequence would come sufficiently far towards the present to cross-correlate significantly with the established modern chronology.

The buildings considered in this chapter and the timbers from them mostly came to light through the interest and attention of a number of vernacular architects and local historians. The sites which produced useful timbers consti-tuted only a small fraction of the buildings (mostly houses) investigated. The majority of the old mud-walled, thatched houses in Ulster belong to the late eighteenth or early nineteenth centuries and were built with bog oak or imported timber. A published example, typical of such houses investigated, was the Articlave House, Co. Londonderry (McCourt, 1966). This supposedly eighteenth-century house of cruck-truss construction yielded six timbers. These were all primary, consisting of three wall posts and three cruck blades. Of the six samples, four were of bog pine, one of bog oak and the other of re-used forest oak. This assemblage of poor-quality material came from a building which was situated within a mile of the Liffock house, cited below, which was built entirely using forest oak.

## Hillsborough Fort

The Fort at Hillsborough is believed to have been built either before or during the 1641 Rebellion (HMSO, 1940). The earlier estimate is probably correct considering the state of the Province during the Rebellion. Unquestionably it is a seventeenth-century star-shaped fort situated on a rise slightly to the east of the town of Hillsborough, Co. Down. In its hey-day it controlled the important route from Dublin to Carrickfergus, the Pass of Kilwarlin. Its importance can be gauged by the fact that it was made a Royal Fort by Charles II and was visited by William III on his way south to do battle with James II in 1690.

Structurally, the fort consists of four walls with spear-shaped bastions. In the centre of the west wall is an enlarged gatehouse known as the 'castle' (see Plate 5(a)). It was from the castle that a group of 25 oak floor joists were recovered

**Plate 5: (a) The enlarged gate house of Hillsborough Fort, Co. Down, which yielded the first large group of seventeenth-century timbers.**

during renovation work in 1968. At first sight the beams could have belonged to the seventeenth-century phase of building. However, two points argued against this. First, the beams, on removal, exhibited extensive signs of re-use. Numerous slots, pegged mortices and angled joints showed that their original function had been something other than the carrying of floor-boards. Second, it is known that the gatehouse was re-furbished in the mid-eighteenth century.

So when first obtained there appeared to be three possibilities for the dating of the beams. They could be eighteenth-century, or re-used from some seventeenth-century portion of the Fort or re-used from some other external source. A reasonable case can be made for the third possibility and it is worth some elaboration, as it demonstrates the complex history of what, on the surface, might have been a straightforward group of eighteenth-century timbers.

A Mrs Delaney, wife of the Dean of Down, visited Hillsborough on 1 October 1758 and left the following record (Bentley, 1861):

to a castle that Lord H. is building. The old castle is fallen to decay, but as it

**Plate 5: (b) Gloverstown House (originally from Co. Antrim) at the Ulster Folk Museum, Cultra, Co. Down. This was the latest building, so far discovered, to have been built with oak timbers in Ulster (AD 1716).**

is a testimony to the antiquity of his family he is determined to keep it up. The castle consists of one very large room with small ones in the turrets; the court behind it measures just an English acre . . . When this is finished he proceeds to the building of his home [now Government House] , which is to be magnificent and in a finer situation than the one he at present inhabits, and about a mile from it; the castle will stand between them. And what do you think this magnificent man means to do with his present dwelling, improvements, lake and island? Nothing less than making them a present to the Bishopric of Down!

The relevance of this information will be seen when the Fort is considered in relation to the church of St Malachy which is situated only 50 metres to the north. The present church of St Malachy was the third to be built in the area. The earliest

Plantation church, on a different site, was erected in 1636 by Peter Hill. It was destroyed during the Rebellion of 1641. After the restoration in 1660, Arthur Hill erected a new church which was dedicated in 1662. It stood on the site of the present church.

Briefly, when Wills Hill (1718-93), the first Marquis of Downshire, inherited the Hill estates he planned the development of Hillsborough. At that time the county town of Down was Downpatrick. The cathedral at Downpatrick was in ruins and Wills Hill saw the opportunity of usurping both the ecclesiastical and judicial power of Downpatrick to Hillsborough. This is reflected not only in the remark by Mrs Delaney, but is shown in the construction of the courthouse in Hillsborough which was obviously intended for an assize court. In furtherance of this scheme the first Marquis had the church of 1662 pulled down and rebuilt partially on the same foundations. This church, which cost £20,000 and took 13 years to complete, was opened in August 1773. No new service of conse- cration was performed (Barry, 1962, 19) which seems to prove that the earlier church had not been destroyed but merely improved. This church was, and is, a magnificent structure, worthy of cathedral status and even includes a bishop's throne which, despite the efforts of the Marquis, was never used. The seat of power in Down remained resolutely in Downpatrick.

However, from the point of view of a possible date for the beams from Hillsborough Fort, a theory can be evolved. As quoted above, the Fort was being worked on in late 1758. If the church, finished in 1773, took 13 years to build, work on it must have begun around the time of Mrs Delaney's visit to Hillsborough. The timber used in the reconstruction of the Fort could then have come from the demolition of the church of 1662. Strong supporting evidence for this early date for the beams comes from the fact that imported pine was used in the con- struction of all the other eighteenth-century buildings in the town, the local supplies of oak presumably having dried up here, as elsewhere, by 1750.

### The Timbers from the Fort

Among the 25 oak beams there were 5 cases where each of 2 beams had emanated from the same tree. The composition of this group of samples is of interest in itself. All the beams had been produced by sawing. Of the 25 sampled, 20 had derived from different trees and the transverse section of each indicated that the parent trees had been sawn into at least 6 beams and in some cases more. This indicated that the sample of 25 beams came originally from a large total popu- lation of beams, probably greater than 200. This lent weight to the theory that in the first instance they came from a large building and were not cut simply to floor the Fort.

Since these timbers constituted the first major group of post-medieval timbers to be examined, they highlighted a number of basic problems. First, only one of the samples exhibited a definite heartwood/sapwood transition. Therefore, no sample contained a complete ring record as far as its felling date. This meant that one of the most useful guides to the contemporaneity of a group of timbers was missing. Apart from the sample with a trace of sapwood, QUB 544, none of the others bore any physical evidence of relative age, i.e. since any number of heartwood rings could be missing, it was impossible to know which of two samples was earlier or later or if any portion of their ring record was contemporary. This led to considerable problems in establishing visual cross-correlations.

Table 5.1 shows the values of correlation coefficient obtained by computer comparison of individual ring curves. Figure 5.1 shows the ring patterns plotted in their correct relative positions together with a site master chronology.

**Table 5.1: Correlation Values between Individual Ring Patterns from Hillsborough Fort**

| QUB | QUB | 't' Value |
|-----|-----|-----------|
| 536 | 544 | 3.53 |
| 537 | 538 | 4.58 |
| 536 | 537 | 4.20 |
| 542 | 537 | 4.48 |
| 553 | 542 | 5.16 |
| 535 | 542 | 4.95 |
| 536 | 538 | 7.03 |
| 547 | 535 | 4.61 |

The resultant Hillsborough chronology of 261 years was undated. Since the absence of sapwood effectively pushed this floating sequence back in time, it was felt unlikely that any significant correlation could be obtained between it and the modern chronology and indeed none could be found. For this reason, an arbitrary year was chosen as 'Datum' and this floating chronology designated Datum -200 to Datum +60 (Baillie, 1973a).

*Coagh House*

The second major group of timbers to be studied came from a house in the village of Coagh, Co. Tyrone. Coagh lies four miles from the west shore of Lough Neagh. The village in its present form was laid out in the mid-eighteenth century by George Conyngham (Lenox-Conyngham, 1947, 29). There is no doubt, however, that its foundations belong to the seventeenth century. The Conyngham family

**Figure 5.1: Hillsborough Fort ring patterns and the resultant site master in %5 Average format.**

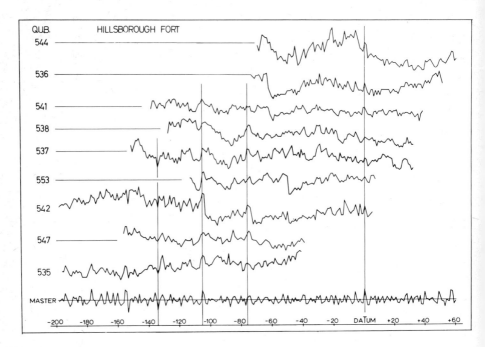

had a mill there at an early date and a letter dated 1719 from William Conyngham of Springhill (see Chapter 6) relates to the building of a bridge at Coagh by public presentment, some thirty years before. In addition, on a map of the Conyngham estates drawn in 1722 (PRO D1449/5, 1) the village of Coagh is shown as containing a number of houses as well as the mill and bridge.

In 1968, extensive renovations to a two-storey house on the north-east corner of the main crossroads uncovered a number of oak roof beams. Strong evidence for an early date for the house could be found in two factors. First, the house occupied a favourable position in the village, at one corner of the main crossroads, about 200 metres from the mill and the river. This crossroads, which is shown on the 1722 map, was the original centre of the village. Second, the mid-eighteenth-century development of the village involved the construction of a very wide street from the crossroads to the mill. Coagh House lies across one end of the wide street, which suggests that the house was already there when the village was improved.

During the renovations to this house in 1968, samples were obtained from 16 oak beams. Of these, 3 were of bog timber and 13 were forest timber. All the beams were of riven oak. As a result, these beams are wedge-shaped in cross-section compared with the rectangular sawn beams from Hillsborough Fort.

Significant cross-correlations were obtained for 6 of these timbers. The remaining samples were unsuitable due to short or distorted ring records. This group as a whole was inferior to those studied from Hillsborough, a fact which points further to the prestige source of the Hillsborough timbers. Of the 6 timbers which cross-dated, QUB 31 exhibited total sapwood, while QUB 26 and 32 both retained portions of their sapwood. The presence of sapwood, combined with the splitting process which normally preserves most of the ring record, made the cross-correlation relatively straightforward. Table 5.2 shows the level of cross-correlation obtained.

**Table 5.2: Correlation Values between Individual Ring Patterns from Coagh.**

| QUB | QUB | 't' Value |
|-----|-----|-----------|
| 26 | 29 | 6.70 |
| 28 | 29 | 5.97 |
| 29 | 39 | 4.70 |
| 31 | 29 | 3.88 |
| 32 | 39 | 5.19 |
| 39 | 28 | 4.17 |

Comparison of the Coagh master chronology with the floating Hillsborough chronology yielded a significant correlation ($t = 5.1$) with an overlap of 92 years. Thus the Coagh chronology covered the period Datum -31 to Datum +116, a significant extension of the floating chronology forward in time. Historically there was good reason to suggest that the two relative felling dates would be about 30 years apart. This assumed that the historical information was broadly correct in assigning dates *c.* 1662 and *c.* 1690 respectively. As this spacing was borne out by the tree-ring match, this in turn suggested that the quasi-historical dating was of the right order. Unfortunately no definitive match could be obtained between the extended floating chronology and the 1649 to 1970 modern chronology.

The search for eighteenth-century oak timbers had to continue in spite of the increasing body of information which suggested that oak had all but disappeared as far as building was concerned by 1725.

*Gloverstown House*

The house at Gloverstown originally stood two miles to the north of Lough Neagh in Co. Antrim. In 1971 it was taken down and removed to the Ulster Folk Museum at Cultra, Co. Down, for re-erection (see Plate 5(b)). The Gloverstown house was a two-storey farmhouse of considerable size and had been thatched up until its removal. With the exception of a nineteenth-century extension, the original fittings were all of forest oak. The most impressive feature was the oak-framed roof of through-purlin, truss construction (Gailey, 1974).

No documentary or verbal evidence for the date of construction of this house was known. On general architectural grounds it could be dated to any part of the century following 1660. Vernacular buildings of this type are basically difficult to date, as parallels with dated English buildings tend to be misleading.

More significant dating evidence could be drawn from the general history of the area in which the house is situated. It was extremely unlikely that the house could have existed before 1660. It lay on the edge of the territory totally ravaged in 1641 and could not have escaped destruction at that time. The use of a simple truss-roof construction suggests that the building was a good deal later than 1660, as most vernacular houses at that time were built using the cruck-truss technique. Camblin's reference (1951, 49) to the desolate state of the western parts of Ulster in 1688 and the associated retreat of the settled inhabitants into defensive towns suggests that the Gloverstown house was probably erected sometime after 1690.

While the Gloverstown house was being rebuilt the opportunity was taken to sample those timbers which were in need of replacement. The timbers sampled all came from the original butt-purlin truss-roof. They fell into two basic categories. First were heavy common-rafters, tie-beams and tie-pieces which were of sawn oak either halved or quartered. In 15 of these heavy timbers investigated only 7 trees were represented. This shows the close-knit grouping of these roof timbers compared with the large original population of timbers which supplied the floor of Hillsborough Fort. Second, there were a large number of light, riven oak rafters. A high proportion of the timbers retained their total sapwood. This was one of the prime considerations in selecting samples (see Plate 2(d)). The only other criterion was that the distribution of samples should be from as many different parts of the roof as possible in order to detect any variation in the age of the timbers or multiple building phases. Correlation of the ring patterns within this group of timber was simplified by the presence of sapwood. Visual correlation was used and Figure 5.2 shows the agreement between six of the samples.

A master chronology of 190 years was constructed for the Gloverstown roof. Cross-correlation of this chronology with the Hillsborough floating chronology

**Figure 5.2: Gloverstown ring patterns with the resultant site chronology linking the Hillsborough and modern chronologies. This link specified the Datum year as AD 1580.**

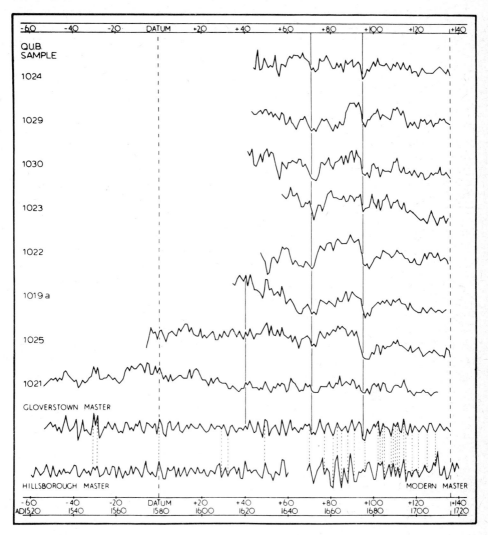

gave a value of $t = 5.8$ with the outer year of the Gloverstown master at Datum +136. In addition, cross-correlation of the Gloverstown master with that from Coagh yielded a value of $t = 7.8$, again with the outer year of the Gloverstown master at Datum +136. The outer years of all but one of the samples, QUB 1021, fall within a two-year range, Datum +134 to Datum +136. The ring record of QUB 1021, which also retained its total sapwood, stopped some years earlier than the rest at Datum +130. However, as this sample was one of the light rafters, it could well have been part of a gathering of convenient light timbers (Baillie, 1974a).

### Cross-correlation of the 336-year Floating Post-medieval Chronology with the Modern Standard Chronology

Due to the small number of very old oak trees available, the modern chronology discussed in Chapter 4 relied for its earliest forty years on the ring record of a single tree. On account of this inherent weakness in the early portion of the modern chronology, the approach to the extension of the sequence back in time was dictated by the available material. This involved the construction forward in time of a floating sequence of tree-rings from post-medieval timbers. None of the structures from which timbers had become available possessed an absolute building date. Thus the floating chronology had a considerable spread of possible dates.

However, by virtue of the cross-correlations between the ring records of the timbers from Hillsborough Fort, Coagh House and Gloverstown House, the relative dates of these three buildings could be estimated. Allowing for the missing sapwood from the Hillsborough timbers, an estimate of the felling date of that group was Datum +76 ± 9. Therefore, the timbers from Coagh House are approximately 40 years later than those from Hillsborough and 20 years earlier than the timbers from Gloverstown House. The linking together of these dates constitutes a significant advance in the available information. On the basis of local history, the timbers from Hillsborough Fort are unlikely to have been felled before 1641 at the earliest. Since the Gloverstown timbers were certainly felled 60 years later, on the basis of their ring records, they must have been felled later than 1700.

With this information available, a cross-correlation was looked for between the floating post-medieval chronology and the modern standard chronology. A significant visual agreement was obtained between these two with the outer year of the floating chronology coinciding with the year 1716. This cross-correlation is shown in Figure 5.2 and produces a percentage agreement of 68 per cent.[2] The computer calculation of a '$t$' value for the cross-correlation was of low significance, presumably because of the considerable fluctuations in the ring record of QUB 528 ($t = 3.2$).

This cross-correlation between the floating Hillsborough/Coagh/Gloverstown chronology and the modern chronology indicated a date for the year Datum as 1580 and established the dates of the Hillsborough and Coagh timbers as 1656 ± 9 and 1696 respectively. These dates fit well with the available historical information and seem to corroborate the theory that the Hillsborough timbers could have come from the church of 1662. The overall chronology was extended to 591 years, from 1970 to 1380. This was the first instance of the construction of an independent tree-ring chronology in the British Isles, involving overlaps (Baillie 1973b).

The only available method for checking this critical cross-correlation, in the absence of timbers whose ring records spanned the same period, was by the dating of timbers from buildings of known date. Chapter 6 looks first at the evidence provided by the timbers from three buildings with narrow date ranges and, second, at evidence which provides absolute certainty of the correctness of the dating.

## Notes

1. It is certain that the oaks were not totally missing. It is likely that many areas of remaining forest were enclosed and hence no longer available for the construction of vernacular houses. The grander residences on the other hand required timber spans so large that they could only be met with imported pine.

2. At the time this dating exercise was carried out percentage agreement coefficients were still in use and CROS was under development. There is, after all, nothing wrong with this matching technique. At Belfast *t* values were normally used for consistency.

# Chapter 6

# Confirmation of the Belfast Chronology

The construction of the continuous 591-year chronology at Belfast was conducted in isolation. It was felt imperative that if a long chronology were to be constructed covering thousands of years (see Chapter 10), the method be proved workable right from the start. Had it proved impossible to link modern to post-medieval trees independently, what hope would have existed for cross-matching all of the sections in the distant past? Other tree-ring projects in England had relied heavily on the suggestion that 'region to region' bridging can form a more elegant means of dating a floating chronology than trying laboriously to link the past with the present' (Fletcher *et al.*, 1974, 39). Lying so much further to the north and west it was not possible to rely on the luxury of such region-to-region bridging at Belfast. This chapter investigates four levels of substantiation of the original 591-year chronology. The information is offered in some detail as a model of the quest for absolute certainty which must underlie the construction of any reference chronology.

### Level 1, Confirmatory Dates

The modern chronology, 1970 to 1649, had been extended to 1380 using the Gloverstown/Coagh/Hillsborough complex. This extension relied solely on an overlap between two ring patterns and appeared to be acceptable. It is essential in this type of study that an air of healthy scepticism should prevail. Let us assume that the overlap and the resultant tree-ring match had been incorrect. What would this have done to the chronology? Would a mismatch between the historic section and the modern chronology have rendered the whole chronology useless? The answer is no, it would not. It must be understood that the only possible weakness in the chronology was the 1649 to 1716 overlap. The two sections Datum -200 to Datum +136 and 1649 to 1970 were each internally robust. The three site chronologies comprising the historic section were so strongly correlated and replicated that no matter what happened to the 1649 to 1716 overlap, the three site chronologies could not be moved relative to one another. The worst that could happen was that the whole historic section would have to move back or forward in time.

However, having made the decision that the 1649 to 1716 overlap was a true tree-ring match, it was important that this be verified. It must be remembered that the historic dating of Gloverstown, Coagh and Hillsborough was vague. It would have been possible to move the dates of these buildings without upsetting any historical evidence; i.e. the dates attributed to these buildings by dendrochronology did not inherently confirm their correctness. An obvious way to check the validity of the chronology was to date timbers from buildings of known age. Unfortunately early buildings in the north of Ireland seldom have precise dates associated with them (see Chapter 5). Examples can, however, be cited where the dendrochronological dates appeared to tie in with known history. Three such examples are discussed below together with a brief résumé of their historical dating.

## Liffock

This National Trust property lies on the north coast of Co. Londonderry, half-way between the villages of Articlave and Downhill. The house is shown in Plate 6(a). Originally it was symmetrical about the front door, the extension to the left being an addition of 1813 (McCourt and Evans, 1973). The principal interest in the house centred on its cruck-truss construction (see Figure 8.1). In attempting to assign a date to the house the following information was used. The house lay to the west of the River Bann and hence was within the area believed to have been totally destroyed in 1641. The same area was recorded as being 'almost desolate, country houses and dwellings burnt' in 1690 (Beckett, 1944).[1] Thus a building date before 1690 seemed unlikely. The Ordnance Survey memoir for Dunboe parish states that the Liffock house was erected in 1691. This survey was made in the 1830s when large amounts of local information, history and folklore were put in writing for the first time[2] (McCourt and Evans, 1973).

When William King was appointed Bishop of Derry in 1690 his first actions included the building and repairing of churches and pressurising of the clergy to make them live in their parishes (Stokes, 1900). Roger Fford, the archdeacon-rector of Dunboe, was certainly resident in the parish by the time of the Bishop's visitation in 1693 (McCourt and Evans, 1973). It is likely that the house was built for Fford at the time of his coming to reside in the parish. McCourt, who researched the house, admitted that there were some contradictions and inconsistencies in the early history, but was none the less convinced that a date *c.* 1691 was implicit.

Although the house was still occupied in the early 1970s, and much of the roof space enclosed, it was possible to obtain samples from two heavy purlins in an original sealed context. These were radial wedges of riven oak, QUB 1014 with

**Figure 6.1: Location of sites within the north of Ireland which yielded late or post-medieval timbers.**

complete and QUB 1015 with partial sapwood. Both samples cross-dated with the Belfast chronology (see Table 6.1).

**Table 6.1: Correlation of the Liffock Timbers with the Belfast Chronology**

| | | | |
|---|---|---|---|
| QUB 1014 | 197 rings | $t = 8.2$ | at 1690 |
| QUB 1015 | 107 rings | $t = 5.6$ | at 1665 |

Plate 6: (a) The 1691 Cruck house at Liffock, Co. Londonderry.

Plate 6: (b) Springhill House at Moneymore, Co. Londonderry, which contained timbers felled in 1697/8.

The relationship between the two dates is of interest with respect to the amounts of sapwood present. QUB 1014 retained all its sapwood, a mean value of 30 rings. QUB 1015 retained only 12 sapwood rings and its outer surface showed clear evidence of axe or edge marks. Since 1015 had its outer ring at 1665, its heartwood/sapwood transition was at 1653 and allowing for missing sapwood its felling date should be in the range 1685 ± 9. This was highly consistent with 1014.

While a larger number of samples would have been desirable to ensure absolute consistency, the agreement between the last year of growth, 1690, and the suggested historical date would appear to be conclusive.

*Springhill House*

Springhill is also the property of the National Trust. It was the family home of the Lennox-Conynghams until the middle of the present century. The house (Plate 6(b)) is situated in the townland of Ballindrum, one mile to the south-east of Moneymore, Co. Londonderry.

There is no historically attested date for the construction of any of the buildings at Springhill. The Lennox-Conyngham Papers, now housed in the Public Record Office in Belfast, which contain detailed correspondence and legal documents relating to the Conyngham family and its property over a 250-year period, scarcely contain mention of the early history. The only two documents dated earlier than 1715 which relate directly to the Springhill Estate are the deed of purchase of the land by William Conyngham in 1658 and the marriage contract between him and Anne Upton in 1680 (PRO D1449/1(5) and D1449/1 (11)), which states: 'the said William intends to erect and build a convenient dwelling house of lime and stone two stories high with necessary office houses'. This indicates that the present house was not built until after 1680. The first letters preserved which are specifically addressed from Springhill are dated 1715. Confirmation of the existence of the house in its present form soon after this date is contained in a book of maps of the Conygham estates dated 1722 (PRO D1449/5(1)). Two plans in this book contain drawings in minute detail of the buildings on the estate and represent the house almost exactly as it looks today. General conditions suggest that if the said William had not built the house by 1685, then it is unlikely that it would have been built before 1690. Lees-Milne (1970) states that

> between 1685 and 1688 the country [England, but no less Ireland] was in a very unsettled state. One pointer to political uncertainty and doubt was the pause in country house building . . . A very significant sequel to the Glorious Revolution was the immediate spate of country houses.

General architectural grounds, in particular the high proportion of roof to front and the absolute symmetry of the front facade (i.e. central door and balanced windows) suggest a date of *c*. 1700 (Cook, 1968). In addition, the height/width ratio of the windows in the original house suggests a William and Mary or Queen Anne date for the central block. While not specific in attributing a date to the construction of Springhill, this varied evidence suggests that the house was built after 1690, within ten years of 1700.

In 1972 the house was investigated for available timber samples. In the roof space and beneath floor-boards riven oak had frequently been used. Of the samples acquired four retained total sapwood. A Springhill master of 10 trees was constructed covering 270 years. This master yielded a correlation value of *t* = 7.8 against the Belfast chronology with its outer year equivalent to 1697. All the samples with complete sapwood ended in the same year. The tree-ring evidence suggests a date which would appear to be highly acceptable for this important house.

## Waringstown Grange

The Grange lies at the centre of the village of Waringstown, Co. Armagh. The village takes its name from its founder, William Waring, who purchased the land in 1658. Waring was responsible for the construction of two important Plantation buildings in the village. The first of these, Waringstown House, was built in 1667, the second, the Parish Church of the Holy Trinity, was built *c*. 1681. Useful as these two buildings would have been in supplying dated timbers, up to the time of writing no opportunity has arisen for the removal of samples. This brings us to the Grange. In 1972 this building was taken over by the National Trust, which undertook extensive renovations; during these the original timbers became available for study.

The Grange stands to the north of the road junction which formed the original nucleus of the village. As late as 1951 it stood at the end of a row of early cruck houses, all of which have now been removed (Camblin, 1951). It is a two-storey house of black stone, its original thatched roof having been replaced with slate in the 1920s. Although not certainly dated, it was considered to be late-seventeenth- or early-eighteenth-century and could be identified with a house on the same site in a map of 1703 (Atkinson, 1934). Although the original roof had been replaced some 50 years ago, it was discovered during survey work in 1972 that the stumps of one of the oak trusses had been left in position when the blades of the truss had been cut away. Apart from these stumps, originally halves of a single tree,[3] three other samples were available. Table 6.2 shows the correlation values between three of the samples and the Belfast chronology. The fourth sample, QUB 1120, although extremely short, showed such an excellent visual correlation with QUB 1119 that it is included. The dates were of considerable interest because QUB 1119, 1120 and 1122 all possessed total sapwood and, in two cases, bark (a rare occurrence). Thus felling dates could be specified and two clear phases demonstrated for the house. QUB 1122 last grew in 1658. Reference to QUB 1121 shows that if missing sapwood is allowed for, an estimated felling date would be 1658 ± 9, which is highly consistent with QUB 1122. The coincidence of 1658 with the known date of purchase of the area by Waring is striking, as is the situation of the Grange at the centre of the village.[4] This is a likely situation for one of the earliest houses. Since QUB 1122 definitely came from the lower floor of the house, it raises the possibility that originally the Grange was a single-storey cruck building like the others in the village. The occurrence of the roof truss of 1692 presumably signifies the raising of the house to two storeys at that time. This could be a house upgraded for one of the weavers brought to Waringstown from, the Low Countries following Samuel Waring's visit there in 1688. These dates also explain why the more important buildings, Waringstown House and the parish church of 1667 and 1681, were well away from the centre of the village. The village was already there when they were built!

**Table 6.2: Correlation of the Grange Timbers with the Belfast Chronology**

| | | | |
|---|---|---|---|
| QUB 1119 | 152 years | $t = 6.3$ | at 1692[a] |
| QUB 1120 | 31 years | visual | at 1692[a] |
| QUB 1121 | 128 years | $t = 4.8$ | at 1626 |
| QUB 1122 | 168 years | $t = 7.8$ | at 1658[a] |

Note a. Total sapwood present

The dates established for these three seventeenth-century Ulster buildings lent strong support to the correctness of the Belfast chronology. Each date fitted well with either the specific or the general historical framework. Of particular importance in this respect were the samples with complete sapwood, as these specified exact felling years.

## Level 2. Cross-dating with England

Further confirmation of the precise positioning of the Belfast chronology was obtained in 1974, when a chronology for the England-Wales border area was made available by Giertz (published as Siebenlist-Kerner, 1978, 157). This chronology covered the period 1341 to 1636 and was dated by direct comparison with the Central German chronologies, with which is showed extremely good agreement. Comparison of this chronology with the Belfast 1380 to 1970 chronology yielded a good visual agreement with $t = 8.1$ at the year 1636. Here then was an independent check: in effect it was possible to 'date' the Giertz chronology against both German and Irish chronologies and arrive at the same date. This result also showed up the stepwise nature of the agreement from Ireland to the Welsh borders to Germany. The two extremes failed to match at a statistically significant level as originally hypothesised.

Thus in 1974 this stepwise cross-dating appeared to be the final proof that the Belfast chronology and the 1649 to 1716 overlap were correct. However, in 1974 a group of timbers had turned up in the bed of the River Lagan just to the south of Belfast. These included massive squared beams 8 metres long and with cross-sections up to 45 cm square. It was immediately assumed that these were parts of an ancient bridge, as they lay only some 15 m downstream from Shaw's Bridge. This existing stone bridge was known to have stood since 1708 and to have replaced an earlier bridge on the same site. Samples with complete sapwood from the heavy oak timbers were dated and their outermost rings were variously 1616 and 1617. Here apparently was an early-seventeenth-century bridge. Some rapid documentary research showed that a Moses Hill owned extensive tracts of land on either side of the Lagan in the early seventeenth century, including 40,000 acres in Co. Antrim and 2,000 acres in Co. Down. His house immediately overlooked the site of the 'bridge'. By coincidence the same Moses Hill (ancestor of the Hills who founded Hillsborough: see Chapter 5) had been created Earl Marshal of all Ulster in the year 1617. The story seemed complete. Then the bombshell — the first documentary mention of a bridge on the site was for one completed by a Captain Shawe in 1655. None of the early-seventeenth-century maps or documents showed a bridge on the site before 1655.

Was it possible that the dating of the timbers was wrong? They certainly matched with the chronology, for example QUB 1987 gave $t = 6.0$ for an overlap of 129 years ending in 1616. Therefore in order to move their date the chronology itself would have to be moved. If they had genuinely been felled in 1655 then the original Hillsborough/Coagh/Gloverstown complex would have to be moved forward by about 38 years. What about the cross-agreement between the Belfast and England-Wales chronologies established above? In order to move the Belfast chronology in time, either the match between these two had to be wrong or the dating of the England-Wales chronology against Germany had to be spurious. All this seemed extremely unlikely but for all anyone knew these tele-connections could have been fortuitous. One interesting sidelight was the question of the historical confirmation derived from Liffock, Springhill and Waringstown above. The only direction in which the chronology could be moved, if indeed it was wrong, was towards the present. This was the only direction in which the dendro-chronological dates for these buildings could be moved. Any attempt to move the dates back in time would have made complete nonsense of the written history. For example, it would have been extremely unlikely for buildings in Waringstown to be constructed before the village was founded in 1658. Moving all the derived dates forward in time would simply have made the dating untidy and would have suggested consistent sampling of secondary or replacement timbers.

All of this is simply to demonstrate the level of doubt which can be induced in a study of this type, compounded by a strong belief that if something can go wrong it will!

## Level 3. Archaeological Interactions

While this heart-searching was going on, there was one important area of interaction with archaeological evidence. This related to material excavated at Joymount, Carrickfergus, Co. Antrim, by the late T.G. Delaney. Carrickfergus still retains remnants of its town wall built originally in 1610 (McSkimin, 1909). Inside the east wall of the town, in the south-east corner of the Joymount site, a pre-existing sixteenth-century town ditch had been excavated in 1974. At one corner of one of the archaeological squares an oak post was discovered. Being an apparently isolated find it remained *in situ* in the baulk, pending the close of the season's excavations. It was, however, earmarked as a prospective 'chronology building sample'. In the final week it was deemed permissible to dig into the section to remove the timber. As luck (or perversity) would have it, the post was the only visible part of an oak revetment (see Plate 7) which had been missed by as little as 5 cm when the square was laid out.

This sudden change in character, from a stray post to a timber structure, warranted immediate excavation. To quote the excavator:

> the red clay which had been laid down to seal the obsolete sixteenth century town ditch extended to the east in a deposit over 1 m thick which abutted a heavy revetment of oak planks retained and braced by timber beams. Although surviving to a height of only 1m this represented the remnant of a clay platform c. 2.50m high which was deliberately slighted soon after construction and sealed by the building level of the 1610 town wall (Delaney, 1974, 7).

The timbers from this structure retained sapwood and dated against the 1380 to 1970 chronology. The outer years of two samples QUB 1595 and 1596 were 1591 and 1601 respectively, suggesting accumulation of timbers to build the revetment in the first years of the seventeenth century.

Archaeologically the revetment and the clay bank it retained were interpreted as 'a temporary gunplatform built to accommodate one or two light pieces

**Plate 7: Revetted gun-platform (?) at Joymount, Carrickfergus, Co. Antrim, sealed by the construction level of the AD 1610 town wall. Dated dendrochronologically to the first decade of the seventeenth century (see also Figure 6.2).**

commanding the east coastal approach pending the completion of the town wall' (Delaney, 1974, 7). So apart from demonstrating the close juxtaposition possible between dendrochronological dates, archaeological remains and historical information, these dates made it difficult to move the chronology forward in time (as possibly indicated by the Shaw's Bridge timbers above). At Joymount the timbers were sealed under a 1610 construction level (see Figure 6.2).

### Level 4. Cross-dating with Scotland

Although the doubts raised by Shaw's Bridge had been allayed by the Joymount exercise (for an example of a similar 'relative dating' exercise which did not turn out so well, see Chapter 12), the author would have preferred a more direct *dendrochronological* proof involving tree-ring overlaps. Fortunately the answer

Figure 6.2: Schematic section through the revetment at Joymount, Carrickfergus, showing its relationship to the town wall of 1610 (see also Plate 7).

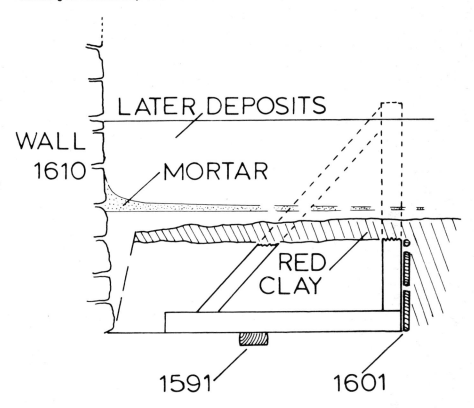

was at hand. In 1976 and 1977 modern Scottish oaks were being sampled (see Chapter 4). These yielded a chronology from 1975 back to 1444, which could not be moved in time under any circumstances, since it was anchored to the present. The ultimate test of the integrity of the Belfast chronology was as follows. The historic section of the chronology, i.e. the Hillsborough/Coagh/ Gloverstown section, formed a replicated unit from Datum -200 to Datum +136, believed to end in 1716. This section of historic chronology was run against the modern Scottish chronology covering 1444 to 1975. The result was a single significant output, $t = 6.2$ with Datum +136 equivalent to 1716.

So there it was. The Belfast chronology 1380 to 1970 confirmed first by dating historic buildings, independently confirmed by stepwise correlation to the Welsh borders and Germany and again independently confirmed by cross-dating with the Scottish modern chronology. This was fortunate, since the chronology had been committed to print as early as 1973 (Baillie, 1973b, 27). Clearly the Shaw's Bridge timbers, whatever they had been part of, had last grown in 1616-17. Both the dendrochronology and the archaeological interpretation of Joymount were vindicated.

This dating substantiation has been laboured in this chapter. It is safe to say that there is no shadow of doubt relating to the date of the chronology. This is an important factor in dendrochronology; the reference chronologies must be precisely correct if they are to be used to provide dates for comparison with historical chronology. Once they are correct, as demonstrated here, the dendro-chronologist is justified in backing his dates against any historical source.

## Notes

1. Liffock lay on the marching route of James's army and could hardly have escaped destruction had it existed before 1690.

2. Unfortunately the same memoir attributed the erection of the house to a Rector Golden, who is not recorded as a rector until 1749. Since the Hazlett family, who occupied the house until the 1970s, took over the house from Golden in 1761, it is probable that his was the only name remembered by the family in 1830.

3. A common trick to ensure symmetry in early timber structures was to make the opposite sides out of halves of a tree, thus ensuring identical length and curvature.

4. One question which arises in the interpretation of a dendrochronological date relates to the interval between felling and use. Fortunately with respect to the study of buildings and archaeological remains seasoning does not appear to be a problem. There is good evidence to suggest that timbers were used green and this is reinforced by the evidence of the tree-ring dates themselves. Hollstein presents a list of some 24 historically dated buildings where in each case the tree-ring evidence suggests use of the timbers either in the year of felling or within a year or so afterwards (Hollstein, 1979, 37). Although there are many fewer historically dated buildings in Ireland it is clear in the case of both Waringstown grange and Liffock house that the timbers were used immediately.

# Chapter 7

# Extension to the Medieval Period

From an archaeological point of view the construction and verification of a reference chronology back to 1380 did little more than set the scene. Late and post-medieval dating applications exist, but the real need for tight chronological markers and horizons lies in the dating of structures of the medieval period and the first millennium AD. In this chapter the problems met with in the further extension of several British Isles chronologies are covered. The basic unit of time is some 500 years, from the fourteenth to the ninth century. This unit of time was not chosen. It became apparent as the work progressed that available timbers from what may broadly be called the Norman period, the eleventh to fourteenth centuries, reached their maximum backward extent in the ninth century.

## Extension of the Belfast Chronology to AD 919

By 1973 overlaps between the ring patterns of modern oaks and those of timbers from Ulster buildings of the eighteenth and seventeenth centuries had allowed the extension of a Belfast chronology to AD 1380. Extensive searching for suitable fourteenth- and fifteenth-century oak timbers brought to light some interesting and worrying factors. In the north of Ireland no buildings with extant timbers survived from before AD 1600. Many seventeenth-century buildings did, however, produce oak timbers and a consistent pattern began to emerge. The oldest of these oaks, felled in the seventeenth century, had *all* started life in the last decades of the fourteenth century. Historically it had always been assumed that the forests present in the north of Ireland at the time of the Plantation were primeval (remnants of the original indigenous forests). The tree-ring evidence appeared here to be at odds with historical considerations. If these building timbers were in fact from ancient forests, it was curious that no examples exhibited ring patterns longer than 280 years, since oaks can reach ages of 400-500 years in natural conditions within the British Isles. The assumption had to be that the forests existing in the seventeenth century were the result of regeneration in the late fourteenth century (see Chapter 11).

The limited extent of these medieval timbers coupled with the lack of early buildings with extant timbers posed a problem for sources of material for further extension of the 1380 chronology in the north of Ireland. Fortunately, sources other than buildings did exist. By 1974 a number of timbers, from a natural source, had yielded a floating chronology spanning the approximate period AD 1000 to 1450. These oaks, dredged from the bed of the River Blackwater in 1969, were sub-fossil or 'bog' oaks. However, two radiocarbon determinations, UB-287 (ad 925 ± 60) and UB-55 (ad 1125 ± 35), placed them in the early Middle Ages (Smith *et al.*, 1971, 1973). Thus they represented one of the youngest groups of naturally preserved oaks encountered during extensive studies within the north of Ireland. The Blackwater trees unfortunately proved to be a limited source of material and only one of them, QUB 51b, extended forward far enough in time to overlap with the Belfast 1380 to 1970 absolute chronology. The ring pattern of this tree showed a significant agreement with the Belfast chronology, $t$ = 3.70, with an overlap of 83 years ending in AD 1462. At that time early medieval dating was in its infancy and it was felt inadvisable to base a master chronology on the ring pattern of a single tree at that point. It was clear that more material was necessary to bridge the fourteenth century and confirm the cross-agreement for the period AD 1380 to 1462.

To place this work in context, the level of isolation under which it was conducted has to be understood. The closest absolute chronologies were those of Huber and Hollstein in Germany, both running back to the ninth century. In England an art-historical chronology had been published for the period 1230 to 1546 (Fletcher *et al.*, 1974, 32). However, this chronology showed no positive agreement with either that for the England-Wales border area or the Belfast chronology (Baillie, 1978, 35). Neither did it cross-date against the Huber/ Hollstein chronologies. On this basis it seemed unlikely that the art-historical chronology would help in confirming the Blackwater extension. Since further local timbers seemed to represent the only way forward, other potential sources were investigated within the north of Ireland.

The solution to the problem ultimately came from a group of crannogs. These are artificial islands or lake dwellings, a number of which were either built or refurbished in the fifteenth or sixteenth centuries (see Appendix 1). As a group they are relatively unexcavated. Most information relating to their periods of use had been gained from assemblages of archaeological remains, mostly accumulated by collectors in the nineteenth century. Between 1969 and 1976 several crannogs were visited when water levels were low. A few were found to have accessible structural timbers preserved by virtue of their waterlogged condition. Ring patterns covering the fifteenth century were obtained, but only one timber, QUB 968 from Lough Eyes, Co. Fermanagh, covered part of the fourteenth century. In 1977 an extensive programme was undertaken by members of the

Archaeological Survey Section of the Department of Environment for Northern Ireland. This survey involved visiting over 100 crannog sites in Co. Fermanagh, and since the establishment of construction dates constituted one of the major aims of the programme, dendrochronologists were involved from the beginning. In the course of this co-operative work some of the sites with accessible timbers were visited in the late summer of 1977 and samples removed for dating. One of these sites, Corban Lough, provided a series of oak timbers felled in the mid-fifteenth century. These timbers provided ring patterns which spanned the fourteenth century and confirmed the Blackwater-Belfast cross-dating. The Corban Lough timber which yielded the longest overlap with the Belfast 1380 to 1970 chronology was QUB 3017, the *t* value being 3.69 for an overlap of 77 years. Both QUB 51b and QUB 3017 were tested against the Belfast chronology at every position of overlap and no other significant cross-correlations were found. In addition, the cross-agreement both visually and statistically (*t* = 4.7) between these two samples was highly significant at the position specified by the individual agreements with the 1380 to 1970 chronology.

One further 'natural' timber, QUB 942, originally thought to be bog oak, from Toomebridge, Co. Antrim, had been dated by radiocarbon to the medieval period (UB-993, 405 ± 40 bp, Pearson, unpublished). This timber also confirmed the Belfast extension. However, simultaneously with the Corban Lough find, a further series of Toomebridge timbers was obtained. This group yielded a 270-year

**Figure 7.1: Closed circle of agreement between the ring patterns which extend the Belfast 1380 chronology. Each matches at a self-consistent position with all the others.**

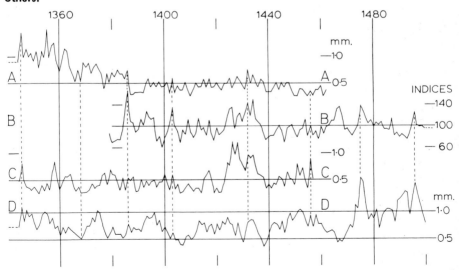

chronology which cross-matched with the 1380 to 1970 chronology and spanned the years 1231 to 1500. This cross-agreement is shown in Figure 7.1 and the respective statistical correlations are listed in Table 7.1 (Baillie, 1977b).

On the basis of this corroborative evidence an extended Belfast chronology was constructed using ring patterns from the Blackwater River, crannogs in Lough Eyes, Mill Lough and Corban Lough and natural timbers from Toomebridge. The extended Belfast chronology covered the period 1970 to 919. Its construction relied solely upon replicated cross-matches between tree-ring patterns and was therefore independent of any historical or archaeological considerations.[1] Dates established by cross-matching ring patterns against this chronology would therefore be independent dates and could be used to check written or otherwise dated sources. The consolidation of this chronology had immediate implications with regard to the dating of a pre-existing Dublin chronology based on timbers from the extensive medieval excavations in that city during the 1970s.

**Table 7.1: Statistical Comparison of the Ring Patterns in Figure 7.1**

| Comparison | *t* Value | Years Overlap |
|---|---|---|
| B cf. A | 3.7 | 83 |
| B cf. C | 3.7 | 77 |
| B cf. D | 5.2 | 121 |
| A cf. C | 4.7 | 124 |
| A cf. D | 5.0 | 129 |
| C cf. D | 4.8 | 132 |

A = Blackwater QUB 51b
B = Belfast 1380-1970 index chronology
C = Corban Lough QUB 3017
D = Toomebridge mean chronology

## Dublin Medieval Dendrochronology

Between 1969 and 1976 extensive excavations were carried out on sites in High Street, Winetavern Street and Christchurch Place within the medieval city of Dublin and at Woodquay which lies between the medieval city wall and the River Liffey (Figure 7.2(a)). These sites contained extensive waterlogged organic deposits in which timbers and timber structures were frequently preserved along with outstanding archaeological remains reflecting almost every aspect of medieval life (ORíordáin, 1971). In general terms the sites could be divided into an earlier

Figure 7.2: Location of medieval sites yielding oak timbers: (a) in the area of the Belfast study 1. Toomebridge; 2. Blackwater; 3. Lough Eyes; 4. Corban Lough; 5. Mill Lough; (b) in the Dublin area 1. Christchurch Cathedral; 2. High St.; 3. Christchurch Pl.; 4. Winetavern St.; 5. Woodquay; (c) in south and central Scotland.

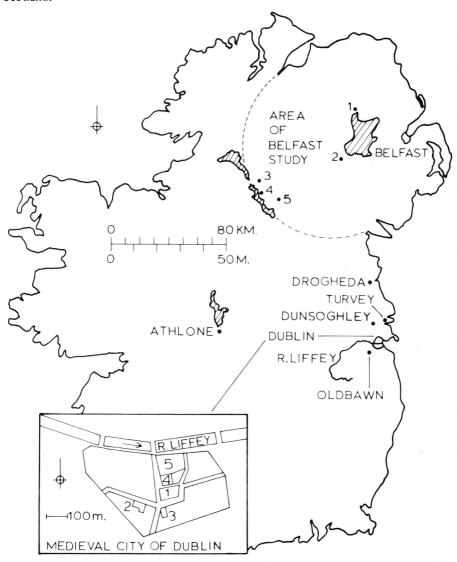

**Figure 7.2: (c)**

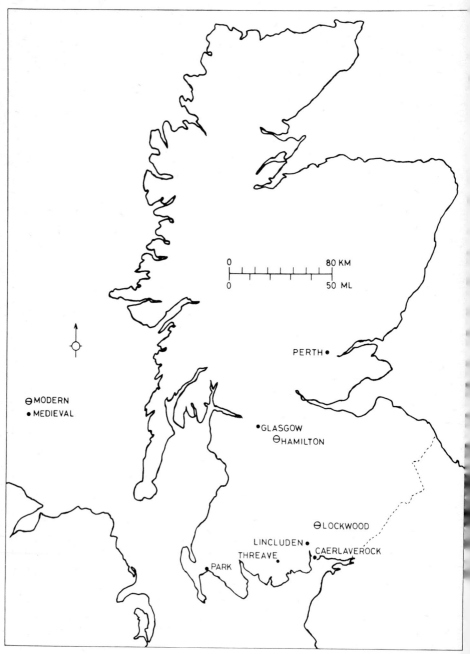

tenth- to twelfth-century phase when the predominant heavy timber was ash and a later twelfth- to fourteenth-century phase when the most widely used timber was oak. Those two phases were more or less coincident with the Viking and Norman periods respectively. The dendrochronological work discussed below was restricted to oak timbers and hence mostly to structures of the Norman period.

In order to reduce problems to a minimum, only timbers which showed no signs of having been used in ship construction were accepted. This involved rejection of planks with rows of iron nails or perforations as well as those with chamfers and notches unrelated to the structure in which they occurred. In this way it was hoped to avoid the wasteful procedure of attempting to cross-correlate foreign timbers.[2] Although this policy was adhered to during the construction of the Dublin chronology, it may well have been a needless precaution. Subsequent analysis has shown that most boat- or ship-derived timbers from the Dublin excavations cross-matched with the Dublin chronology with correlation levels sufficiently high to suggest local origin (Baille, 1978b, 259; see also Chapter 12).

Before the onset of the archaeological excavations on sites such as High Street, the overlying Georgian houses and cellars were removed. Immediately beneath the cellar floors were deposits of the thirteenth century. All of the later levels relating to the period between the thirteenth and the eighteenth centuries, if they had ever existed, were missing. It was possible at that stage to formulate a problem for solution. Timber structures lying on the surface of the archaeological levels or cutting through the upper levels could be either thirteenth-century or they could possibly be later structures dug in from higher levels and truncated at the thirteenth century. The primary aim of the initial tree-ring project was the dating of these various timber structures, mostly pits or house frames.

The first step had to be the construction of a floating chronology for the sites. Subsequent timbers could then be dated relatively and hopefully the chronology could be extended and dated. The floating chronology was based on two principal groups of timbers derived from a plank floor and house frames TFS 2 (Timber Framed Structure) and TFS 3 respectively. Stratigraphically, these structures, excavated at High Street, bore a close relationship in that TFS 2 was overlain by a later post and wattle wall which was subsequently cut through by TFS 3. When the ring patterns from the individual timbers of these structures were cross-matched and submasters produced, it was found that TFS 2 pre-dated TFS 3 by 106 years (Baillie, 1977c). The cross-correlation between the submasters gave a value of $t = 11.1$ for an overlap of 154 years.

In order to facilitate reference to the floating chronology an arbitrary ring was chosen as Dublin Datum (DD) and the 391-year TFS 2/TFS 3 chronology labelled DD -213 to DD +177. Subsequent cross-dating of a series of timbers from other structures as well as miscellaneous long-lived timbers succeeded in extending the floating chronology back to DD -274. All the pits and house frames studied cross-matched with the resultant 452-year chronology. No timbers were found which extended the chronology forward in time, and thus TFS 3 was the latest structure on any of the sites investigated up to 1973.

The major question in 1973 was the absolute dating of the Dublin chronology. As a floating chronology it was of use in assigning relative dates to a number of important pits, houses, wharfs, etc. However, the importance of some of the associated groups of objects, from an archaeological point of view, demanded precise dating. Various strands of evidence, including a cross-agreement ($t = 4.14$) with the then unconsolidated extension of the Belfast chronology (Baillie, 1977b) suggested that the Dublin 452-year chronology started in the year AD 855. It was felt, in 1973, that a desirable approach would be to complete an independent Dublin area chronology: that is, to extend the tentatively dated floating chronology forward in time to cross-match with material of known date.

Extensive sampling of timbers on all the Dublin sites over several excavation seasons had failed to produce any ring patterns which extended the 452-year chronology forward in time. The vast majority of oak timbers studied belonged to the thirteenth century. One group of posts, however, from high up in an unexcavated baulk at Christchurch Place, did yield a 200-year ring pattern which could not be cross-dated with the existing Dublin chronology. It was assumed that this 200-year chronology had to be later than the 452-year chronology and this was confirmed when cross-dating was obtained between the baulk timbers and a group of four ring patterns from Bathe House, Drogheda. Bathe House was dated to 1570 by an inscription and since one of the timbers exhibited a heart-wood/sapwood transition it could be inferred that the 200-year ring pattern ended in the sixteenth century (see Chapter 8).

Subsequently a 200-year floating medieval chronology was produced from these timbers and from further samples from Turvey Castle and Oldbawn House, both Co. Dublin, and Christchurch Cathedral, Dublin. This chronology exhibited strong cross-agreement with the chronologies from Belfast and the England-Wales border area (see Figure 7.3). In both cases the 200-year Dublin chronology ended in the year 1556, the cross-agreement values being $t = 7.1$ and 6.0 respectively. Thus this later Dublin chronology spanned the years AD 1357 to 1556. With the exception of the single timber from Christchurch Cathedral, each of these sources of later medieval timbers yielded at least one example which started life in the second half of the fourteenth century. No examples were found which started life earlier. This appeared to support the finding from the north of Ireland that there was a depletion/regeneration phase for oaks at that time.

The problems posed by this hypothesised depletion/regeneration phase can be elaborated by an example. It would appear that the way to bridge the fourteenth century should be to acquire timbers of the fifteenth century. Now there are very few timbered buildings of the fifteenth century in Ireland. One which did

**Figure 7.3: Cross-correlations between the Belfast, Dublin and Giertz chronologies.**

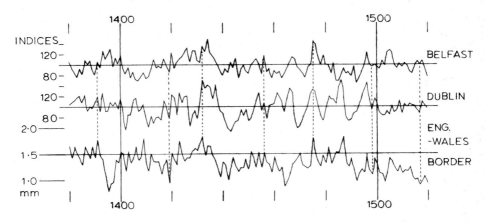

exist and which should have been ideally placed to bridge this Dublin area gap was Dunsoghley Castle, Co. Dublin. Dunsoghley is reputedly the last medieval castle in Ireland to retain its original timber roof. Historically its building date is not that clear, but it is unanimously held to belong to the first half of the fifteenth century (DeBreffny and Mott, 1977, 110). Examination of the *in situ* roof trusses and coring of a truss removed during renovation showed that the timbers used came from young oaks, none of the cores yielding more than 40 rings. Although these short ring patterns could not be dated dendrochronologically, they pointed to the timbers having come from trees regenerating in the late fourteenth century.

No further progress was made with the extension of the Dublin chronology or the bridging of the fourteenth century until 1977, when consolidation of the Belfast chronology back to 919 confirmed the placement of the Dublin chronology as running from 855 to 1306 (Baillie, 1977b). The Dublin chronology consists not of one continuous chronology, but of two precisely dated sections. While it proved impossible to find timbers to cross the fourteenth century, historical considerations suggest that it would be extremely difficult to acquire modern oaks in the area which would extend back before the eighteenth century. This situation is due to over-exploitation in the seventeenth century. It is unlikely that a continuous Dublin chronology could be completed. Fortunately the two sections are sufficient to allow dating of most relevant timbers.

## Scottish Medieval Dendrochronology

In Chapter 4 it was noted that long-lived oaks were available for study in south and central Scotland. In particular the modern chronology from the Cadzow forest at Hamilton had yielded ring patterns running back to 1444. This modern chronology was subsequently used to confirm the dating of the 1716 to 1380 section of the Belfast chronology (see Chapter 6). The flying start offered by these very long-lived trees made it logical to attempt the construction of at least an outline chronology for Scottish oaks covering the last millennium. The real impetus came with the publication in 1975 of the bridge timbers from the moat at Caerlaverock Castle near Dumfries (Rigold, 1975, 72). If a suitable chronology could be put together, then it might be possible to date the various phases of construction at this important Border castle (see Chapter 8).

Using the lessons learned at Belfast, the following proposition was addressed to archaeologists and architectural historians in Scotland. If they could supply suitable samples from medieval buildings and excavations, we would build the chronology. The requirements were relatively simple: all they had to do was supply timbers dating from each of the centuries from the sixteenth to the thirteenth. As fortune would have it, contact was made with exactly the right person, Chris Tabraham of the Property Services Agency of the Department of the Environment for Scotland. Under his auspices and with the assistance of Alf Truckle, the Keeper of Dumfries Museum, a single trip to Scotland in 1976 yielded timbers of the thirteenth, fourteenth, fifteenth and sixteenth centuries. Although additional timbers were later necessary to bolster the chronology, the result of that trip was the consolidation of a chronology for southern-central Scotland spanning 946 to 1975. It must be noted that the construction of the equivalent Belfast chronology had taken almost a decade. One important factor in the rapid progress with the Scottish material was the pre-existence of the chronologies for Belfast, Dublin and the England-Wales borders. Once a site master chronology was produced for a group of Scottish timbers, it could be compared not only with relevant Scottish masters but with these pre-existing chronologies.

### Sixteenth Century

For the purposes of chronology building it was necessary to obtain timbers of the sixteenth century. Castle of Park, Kircudbright, had recently been extensively renovated by the Department of the Environment. The castle, a late tower-house, was historically well dated. A carved stone panel over the entrance gave the day of the month in the year AD.1590 when building had commenced. The renovations to the building entailed the removal of the main floor beams and replacement of those whose ends had deteriorated. Samples of five timbers were obtained and the ring patterns of these were found to agree. Unfortunately none of the samples retained its sapwood and only one, QUB 2923, had a clear heart-wood/sapwood boundary. The 202-year Park chronology cross-dated extremely well with the Belfast, Dublin and England-Wales Border chronologies with correlation values $t = 7.1, 9.4$ and $6.8$ respectively, ending in the year 1551. The cross-agreement with the Scottish modern chronology back to 1444 is not of this calibre, but for half its length depends on the ring pattern of a single tree, QUB 2818. It will be noted that the Park chronology extended back from 1551 to 1350. Here again it seemed was the regeneration noted from the mid-fourteenth century at Belfast and Dublin (above). It was realised that if the 1350 'gap' existed in Scotland, it might prove as impossible to complete the Scottish chronology as had been the case in the Dublin area. Fortunately the next group of timbers from Lincluden resolved this potential problem for Scotland at least.

## Fifteenth Century

A single, radially split, oak panel, originally with painted decoration, is housed in the Dumfries Museum. It was known to have come from a choir stall from Lincluden College, Dumfries. The remainder of the stall is housed in the National Museum of Antiquities of Scotland in Edinburgh and was believed to be of fifteenth-century date on stylistic grounds (Baillie, 1978c, 257).

It was possible to extract the ring pattern from a wedge removed from the non-painted surface of the Dumfries Museum panel. This yielded a ring pattern, QUB 2928, which spanned 367 years, apparently ending in AD 1434 on the basis of a tentative overlap with the Castle of Park chronology. Because of the import-ance of timbers of this date range, spanning the fourteenth century, a visit was made to the National Museum early in 1977. The Lincluden stall was dismantled and the polished ends of five further panels were photographically recorded. It was possible to measure the ring patterns of four of these from the resultant photographs. Each of the panels cross-dated with QUB 2928, and confirmed the cross-agreement with Castle of Park, the outer rings being AD 1434, 1452, 1456 and 1467. The Lincluden master chronology gave a correlation value of $t = 3.1$ against the Park chronology with its outer year at 1467. When compared with the Belfast 919 to 1970 and Dublin 1357 to 1556 chronologies, the cross-agree-ments were $t=5.7$ and $6.8$ respectively. In both cases the Lincluden chronology ended in 1467. Thus the Lincluden chronology spanned the years 1068 to 1467.

In addition the Lincluden chronology cross-matched with the Dublin 855 to 1306 chronology, giving $t = 4.9$ for an overlap of 239 years. This confirmed the relative positions in time of the Dublin early and late medieval chronologies previously fixed on separate evidence.

**Figure 7.4: Site units comprising the Scottish AD 946 to 1975 chronology.**

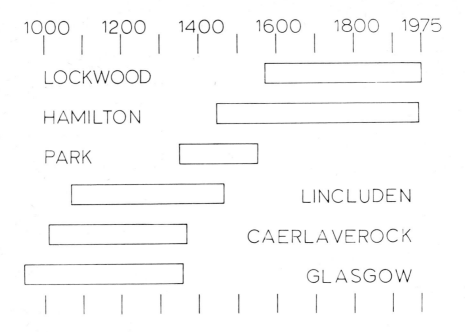

### Fourteenth and Thirteenth Centuries

An important group of timbers from Glasgow Cathedral allowed the consolidation and further extension of the Lincluden chronology. The roof of the Cathedral was extensively repaired in the early years of this century (Oldrieve, 1916). At that time a series of structural beams was removed and subsequently stored in Newark Castle, Greenock. Samples were removed from 20 of these timbers, of which 14 turned out to be of use. The individual timbers cross-matched to give a 415-year chronology containing timbers of two clear phases, one of the fourteenth and one of the thirteenth century. This Glasgow chronology cross-dated with the Lincluden chronology with $t = 8.5$ at an overlap of 293 years ending in 1360. Thus the Glasgow chronology covered the period 946 to 1360. One of the timbers, QUB 2648, extends back to 896, but a series of narrow rings between 920 and 940 rendered this extension unsuitable for inclusion in the master chronology. Figure 7.4 shows the basic units used in the construction of the Scottish chronology (Baillie, 1977a).

**Integrity of the Medieval Chronologies**

In Chapter 6 the ultimate integrity of the Belfast 1380 to 1970 chronology was demonstrated by the replication of the dating in several directions. Can a similar demonstration be applied to the medieval sections of chronology covering the fourteenth to the ninth or tenth centuries for Dublin, Scotland or Belfast? Again the answer is in the affirmative. Figure 7.5 shows the consistent cross-matching between the various sections of chronology from around the Irish Sea basin.

In 1977 a section of chronology for southern and eastern England was published for the period 780 to 1193 (Fletcher, 1977). This chronology was named Ref 6 and was placed in time by cross-correlation with the German chronologies. For example Ref 6 gave $t = 7.5$ when compared with the Hollstein (1965) chronology. When Ref 6 is compared with the Dublin 855 to 1306 chronology the agreement with the outer year of Ref 6 at 1193 is $t = 6.6$. In short, it is possible to date this section of English chronology identically against either Ireland or Germany. This represents the ultimate check on the correctness of the whole complex of medieval chronologies.[3]

Figure 7.5: Cross-correlations linking the independent Irish Sea basin chronologies stepwise to Germany.

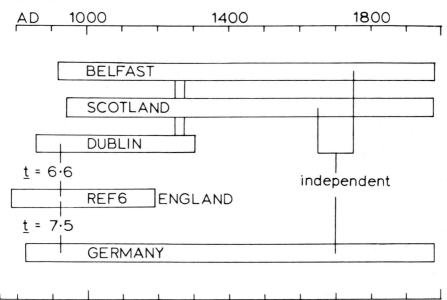

One question which remains relates to the areas within which these various chronologies will allow dating. This is very much an empirical question, but sufficient results are available to allow at least a partial answer to be advanced as follows. It is almost certain that there are now sufficient medieval chronologies available to allow the dating of any new site chronologies within the British Isles. If we consider the unit of time comprised by the tenth to thirteenth centuries AD, there are already chronologies available for the areas broadly localised by the following cities, namely Perth, Glasgow, Dumfries, Belfast, Dublin, York, Liverpool, Exeter and London. It appears likely that any timbers, especially groups of timbers, of this period should be datable against one or other of these indigenous chronologies and/or against one of the German chronologies.

## Notes

1. It is worth drawing attention to the existence of a closed circle of agreement here. This configuration, where a series of ring patterns all cross-agree at self-consistent positions, gives extremely powerful confirmation that the dating is correct.

2. It was not that re-used timbers had to be foreign, but simply that if re-used, especially for ships or boats, the *possibility* existed that they might be foreign. Even this would be sufficient to introduce an imponderable into the chronology construction. Hence the rejection.

3. It must be stressed that this final check does not mean that the Irish chronology depends in any way on the German chronology. Hopefully it has been demonstrated that the Irish/Scottish chronologies are independent of any considerations other than tree-ring matches.

# Chapter 8

# Medieval Dating Examples

Having established chronologies covering the last millennium, they could immediately be used to solve a variety of dating problems. It seems essential to review some examples of dating to demonstrate both the refinement possible with dendrochronology and the limitations imposed on the method by inadequate samples. In point of fact some of the best examples were dated during the chronology-building process and no excuse is made for using these to elucidate various points. In particular two of the sites dated were not chosen because of any dating problem, but were studied for the very reason that their historical dates were known and it was assumed that they would constitute potentially useful sources of ring patterns for chronology building. That in no way detracts from the lessons to be learned from these dating exercises. On the contrary, they give a true understanding of how dates derived from dendrochronological analysis compare with known building dates (similar results comparing derived dates with known dates of panel paintings are touched upon in Chapter 12). The examples chosen range from the seventeenth to the eleventh centuries and are mainly drawn from work carried out at Belfast.

## Standing Buildings

### Seacash Cruck House

In Chapter 6, timbers from three partially documented Ulster buildings were dated. One of these, the Liffock house, was a substantial cruck-framed building whose suggested documentary date of 1691 was supported by a tree-ring date of 1690. Crucks are rare in Ireland compared with Britain. Only some 50 examples have so far been recorded in the country (Gailey and McCourt, 1978). Thus, whenever possible, it is of interest to see when the technique was in use. Reference to Figure 8.1 shows the basic principle of the cruck. Opposing uprights rising from ground level are joined at the apex with further horizontal tie-beams linking the blades at a lower level. Each cruck therefore has an 'A' profile and a series of these crucks are linked together with purlins and a ridge beam. The essential feature of a cruck house is the capacity of this wooden framework to support

**Figure 8.1: Dated crucks in Ulster.**

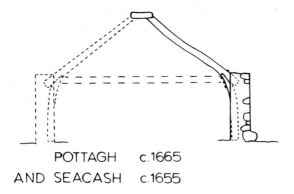

POTTAGH      c.1665

AND  SEACASH      c.1655

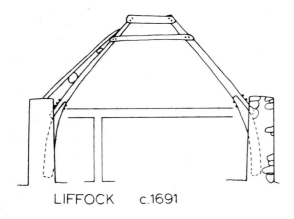

LIFFOCK      c.1691

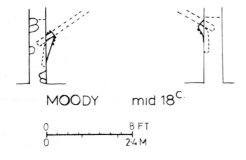

MOODY      mid 18$^{C.}$

the weight of the roof directly, without the necessity for load-bearing walls. The tie-beams serve to counter the outward thrust on the blades due to the weight of the roof. Such a structure can be erected and the walls and gables added later. Frequently the cruck blades are actually incorporated in the body of the walls, clearly demonstrating the order of building (Baillie, 1976).

Originally the technique, as seen in England, employed the two matched halves of curved trees as the blades of the crucks. Specimens of this English type of cruck are known in Ulster, for example at Waringstown where they presumably date (English-derived) to 1658 or after (see Chapter 6). The more common cruck type in Ulster appears to be the composite truss. In this type the rising timbers are made in two sections, a wall post and an angled blade. The two parts are mated with scarfed and pegged joints. Such an arrangement is shown in Figure 8.1. It has to be assumed that this composite-truss technique was a response to a lack of suitable long, curved timbers. There are hints, from documentary sources, that this type of truss was known in Ulster in the sixteenth century, but no example of such an early date has been proven.

Clearly dendrochronology offered an opportunity to investigate the period of usage of this technique in Ireland. In 1972 the ruins of a house containing one half of a composite-cruck-truss were discovered in Pottagh Td., Co. Londonderry (see Figure 8.1). A number of the original timbers were lying within the ruin, including a yoke which had originally linked the apex of the truss and a number of purlins. Of the samples dated, two with complete sapwood, QUB 934 and 940, had end years of 1664 and 1665 respectively ($t$ = 5.2 and 6.5). At the time this was the earliest certainly dated cruck house in Ireland (Baillie, 1974b, 20).

In 1977 renovations to what appeared to be a typical late-eighteenth-century farmhouse at Seacash Td., Co. Antrim, turned up an almost complete composite cruck-truss. This comprised one wall-post and both blades together with a connecting collar. The surviving wall-post was embedded in the stone wall down to ground level. Initially three samples were taken for dating purposes. However the two blades had, as expected, been cut from the same tree, so the samples reduced to QUB 4303 representing the blades and 4302 the wall-post. On measurement 4303 yielded a ring pattern of 168 years ending in 1613 ($t$ = 6.0). This sample had two rings of sapwood remaining so an estimated felling date should have been in the range 1643 ± 9. This appeared to be backed up when the 105-year pattern of QUB 4302 ended in 1610 ($t$ = 7.1) at its heartwood/sapwood boundary, again suggesting felling within 1642 ± 9.

Unfortunately this suggested dating, either just before or just after 1641, posed a problem. Reference to Chapter 5 shows that a date in the decade after 1641 was a fairly unlikely time for anyone to be building in the area.[1] If, on the other hand, the date related to the 1630s, it would have given the first clear indication of the use of the cruck technique in Ireland before 1641. Further, as the

Seacash example was essentially identical in every detail to the Pottagh cruck above, it would have shown a useful continuity. It became important therefore to resolve this dating question.

As renovations continued on the Seacash house the complete wall-post became available. Some extremely worm-eaten sapwood survived down one corner of the squared timber. By slicing this repeatedly a spot was found where the complete sapwood ring pattern could be measured. This showed that QUB 4302 had a total of 45 sapwood rings plus the spring vessels of a 46th ring, strongly suggestive of felling in the early summer of 1655.

Remembering that the wall-post has got to be one of the original timbers, this result seems to settle the date of the Seacash cruck. While still the earliest dated example, 1655 places it in a similar context to Waringstown, where settlement was taking place in 1658. At Waringstown, Waring had purchased the land from soldiers who had been paid with grants of land to encourage them to settle in the area. Probably the Seacash cruck reflects something similar. From the dendrochronological point of view this example serves as a warning about relying on sapwood estimates. Sapwood numbers are highly variable, and the ± 9 quoted on the Belfast estimate approximates to one standard deviation. In that light the Seacash sapwood falls within two standard deviations and hence is perfectly acceptable. Overall this dating exercise gave an initial date about which historians might have been uneasy. Subsequent investigation showed that in this instance the sapwood estimate used, despite being the highest figure in northern Europe, was still inadequate. This points strongly to the fact that a sample with complete sapwood is worth much more than any estimate for making sense of historic dating problems.

### Castle of Park

In Chapter 7, the timbers from Castle of Park, Kirkcudbrightshire, were mentioned as a source of Scottish ring patterns for the middle ages. The five timbers from this site are represented schematically in Figure 8.2(a). It is clear that, allowing 32 ± 9 sapwood rings for QUB 2923, the one sample which showed a definite heartwood/sapwood boundary, we achieve an estimated felling date of 1583 ± 9. If the inscription on the building is correct, construction was commenced in 1590. So while the estimate is adequate, it appears to be somewhat too low and again points to the inadequacy of any estimate in comparison with the presence of total sapwood. For example, if a felling date could have been established, it would have been possible to tell whether the timber was cut in advance of construction or contemporary with it.

### Bathe House, Drogheda

This house, which stood at the junction of Lawrences Street and Shop Street,

Figure 8.2: (a) Relative placement of the Castle of Park ring patterns. (b) Relative placement of the Bathe House ring patterns. (c) Relative placement of the Glasgow Cathedral ring patterns showing two clear phases.

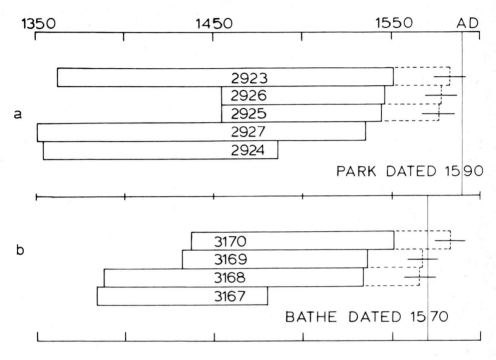

Drogheda (30 miles to the north of Dublin) until 1824, was probably the last timbered Tudor house in Ireland. Its front carried a carved documentary inscription to the effect that the house was erected by one Nicholas Bathe in the year AD 1570. Fortunately these decorated beams survive in the National Museum of Ireland in Dublin. The trees used in the house construction were reputedly derived from Mellifont Park (McCracken, 1971, 73). This implicit Irish origin suggested that these timbers might prove a useful source of medieval Irish ring patterns at a time when attempts were being made to tie down the Dublin archaeological chronology (see Chapter 7). The Director of the National Museum kindly gave permission for the beams to be sampled and this was achieved using a Henson dry corer (see Plate 3(b)). When the beams were being cored two tenons were found still pegged in position in mortices in one of the beams. These were also sampled.

The four matching ring patterns are shown schematically in Figure 8.2(b). Only one, QUB 3168, positively showed evidence of a heartwood/sapwood

**Figure 8.2: (c)**

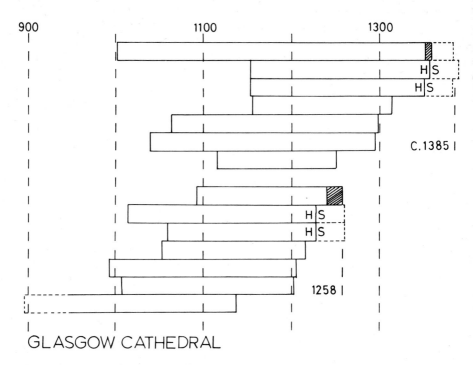

GLASGOW CATHEDRAL

transition. The other beam, 3169, had, however, been sampled to take advantage of a natural indentation on the beam edge where it appeared that only the sapwood was missing. The close coincidence of the end dates of these samples is therefore probably significant. If sapwood is allowed for on these two beams, the estimated felling dates neatly straddle the known building date.[2] However, a question arises when the outermost ring of the second tenon, QUB 3170, is considered. This ring pattern ends in 1551 at a point which may well be in the heartwood/sapwood transition. Allowing sapwood for this sample produces an estimated felling date of 1583 ± 9, erring on the late side. This would appear to be a case where all three samples should be considered as outlined in Chapter 2. These three samples end in what approximates to a linear distribution. In particular they all end within 18 years. It would probably be justifiable simply to take the mean outer year of the three and add the sapwood allowance to that. Doing this gives an average outer year of 1540 and allowing for sapwood suggests felling in the range 1572 ± 9, a date which would certainly be acceptable.

The reader may be inclined to say that the results are being selected to fit the known date. This is not quite the case. It is just possible that we are moving towards formalisation of the 'linear distribution' idea. Obviously it is more realistic to use the information from all the available samples than to select arbitrarily. Here is a case where the end years fulfil the quasi-condition of falling within 18 years. Perhaps the condition could be formalised into a hypothesis: 'if three or more timbers from a structure show end years which lie within eighteen years then a felling phase should have taken place 32 ± 9 years after the mean end year'. Referring back to Chapter 2, the linear distribution condition was suggested for samples without sapwood. If it stands up for such cases it should be perfectly valid for cases where at least some sapwood is present.[3] The difference, however, is that in Chapter 2 it was arbitrarily suggested that five or more samples must end within 18 years to define a phase. Here, however, using the Bathe House timbers, we have reduced to three. Now reference to the results from Glasgow Cathedral, Figure 8.2(c), shows us immediately that we have moved on to dangerous ground if we define a phase every time three timbers end within 18 years. On that criteria there could be four phases of timbers represented in the Glasgow assemblage. Although this is conceivable, it is very unlikely. Therefore the hypothesis does not stand up to even casual scrutiny. Perhaps it has to be changed as follows: 'Where the heartwood/sapwood transition is present on more than one sample, three or more ending within eighteen years specify a building phase. In the total absence of sapwood five or more must end within eighteen years to define a phase.'

This may seem tortuous, but it has to be remembered that this is an intensely empirical subject. Only by observing the behaviour of the data under the controlled conditions of known age groups can it be hoped to formalise an approach which gives some degree of confidence. After all, this discussion surrounds the situation which confronts the dendrochronologist when the precise dating offered by complete samples has been replaced by a much inferior condition.

## Glasgow Cathedral

The fourteen useful timbers from the roof of Glasgow Cathedral (Chapter 7) had been preserved without documentation. They were thus treated as a random group and internal cross-dating was looked for. It became obvious that the two samples which possessed sapwood, QUB 2639 and 2642, did not belong to the same felling phase. These two timbers had been felled around 130 years apart. In each case other samples exhibiting heartwood/sapwood boundaries backed up the groupings (see Figure 8.2(c)). Of particular interest are the two earlier samples QUB 2649 and 2650, which both end in 1228. Allowing for missing sapwood, their estimated felling date would be 1260 ± 9. This is highly consistent with QUB 2642, which had total sapwood and had last grown in 1258.

While showing that a sapwood estimate of around 30 rings seems to be desirable in Scotland as well as Ireland, the results from Glasgow Cathedral are particularly important in that they show clear phasing. This appearance of phases from a random group of timbers is important. It demonstrates clearly that any relative dating information will fall out naturally from a systematic cross-dating exercise.

Returning to the building-phase hypothesis generated above in the discussion of the Bathe House dating, if we apply this to the Glasgow first-phase timbers, the three heartwood/sapwood boundaries which fall within 18 years are those for QUB 2642, 2649 and 2650. For these three the mean outer heartwood ring represents the year 1232 and the estimated felling date would be 1264 ± 9. This would seem to be a reasonable estimate of an actual 1258 felling date. This might all seem very rosy, a hypothesis which allows reasonable estimation of felling dates which seems to hold up in practice. However, it might be worth considering again the timbers from Castle of Park, above. There the estimated felling date of 1583 ± 9 was generated for only one sample, but three timbers, QUB 2927, 2926 and 2923, ended in 1535, 1546 and 1551 respectively (see Figure 8.2(a)). If we were to apply the hypothesis to these, as they fall within 18 years, the mean outer year would be 1544 and the estimated felling date 1576 ± 9. Clearly this is considerably too old for a documentary date of 1590.[4] The single sample estimate of 1583 ± 9 was much better. Does this negate the hypothesis? Well, no, because reference to the latest version above shows that it was cleverly worded to avoid this problem. 'Where the heartwood/sapwood transition is present on *more than one* sample, three or more ending within eighteen years specify a building phase.' Only time will tell if this approach to estimating felling dates for groups of timbers will stand up in practice.

## Archaeological Timbers

### Caerlaverock Castle

This important Border castle is of a rare triangular form surrounded by a wet moat and other defences. It lies some 9 km south of Dumfries and occupies a strategically important position on the Solway coast. At the entrance to the castle the moat was spanned originally by a series of timber bridges, the foundations of which have been preserved. The bridge had been renewed on at least three occasions before it was rendered obsolete by a long drawbridge which spanned the moat to a permanent stone abutment (see Figure 8.3). The moat had been excavated between 1959 and 1966 by MacIvor and details of the bridge construction reported by Rigold (1975, 71). It was this publication which was the prime moving force behind the construction of the Scottish chronology outlined in Chapter 7, viz. if a chronology could be constructed the various building phases might be dated.

The castle itself posed some problems as regards date. It was mentioned in the

**Figure 8.3: Bridge timbers in Caerlaverock moat.**

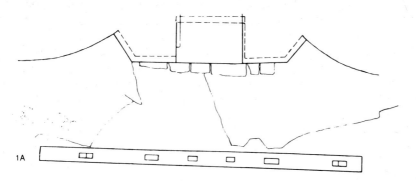

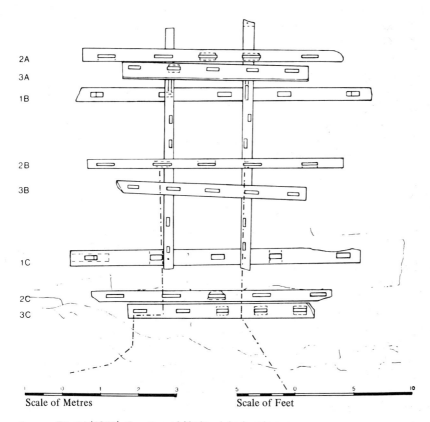

Scale of Metres     Scale of Feet

Source: Rigold (1975). Courtesy of *Medieval Archaeology*.

year 1300 when it was under siege. Although obviously in existence at that time, there was no evidence to suggest when it was built. Since the original bridge timbers existed, it seemed likely that a tree-ring date might clarify the early history of the site. The opportunity afforded by the known multiple phasing of the bridge was also encouraging in terms of elucidating the development of the castle.

To set the scene, MacIvor's interpretation of the bridge history (as published by Rigold) was as follows. The largest timbers formed Phase I and were estimated to be *c.* 1290. Phase II and Phase III replacements both rested on longitudinal timbers which overlay Phase I. These second and third phases were attributed to the second quarter of the fourteenth century and the fifteenth century respectively. A much more ephemeral Phase IV was related to 1593 and to the construction of a rubble abutment which sealed timbers of the first three phases.

Fortunately the timbers had been replaced in the moat by MacIvor in the hope of 'future' dendrochronological dating — admirable foresight. In 1977 the moat was drained by staff of the Department of Environment for Scotland to allow sampling of the timbers. All of the principal timbers of Phases I, II and III could be identified, as could the longitudinals. Unfortunately the Phase IV timber could not be traced. Samples were removed from most of the major timbers and, where possible, attempts were made to ensure the presence of sapwood. Since the beams were squared, sapwood was normally only present on the edges.

Figure 8.4 shows the results of the tree-ring analysis. All the long-lived timbers from Phases I and II could be cross-dated, as could the longitudinals. The failure to date Phase III is discussed below. What was immediately apparent was that three felling dates were represented. The first, 1277 or shortly after, would seem an eminently reasonable time for the construction of Caerlaverock. The date of QUB 2871 is backed up by the estimated felling date of QUB 2854. The interpretation of this date is the province of the archaeologists and architectural historians, but presumably its relevance would depend on whether or not the moat was cut and bridged before the castle itself was built.

Interestingly, the longitudinals, which must have been placed in the moat at the time of the construction of the Phase II bridge, were not the same date as the Phase II timbers. The longitudinals had been felled in the year 1333 and other stray timbers backed up this date. Of particular interest was the fact that in these timbers we had an excellent sample of a Type A1 dating (see Chapter 2). On a number of samples the spring vessels for the year 1333 were apparent, but no summer-wood had formed. This was suggestive of timbers being felled presumably for use somewhere in the castle in May or June 1333. The coincidence of this date with the serious Scottish defeat by Edward at Halidon Hill on 20 July 1333 implies that there was activity at Caerlaverock immediately before the

**Figure 8.4: Relative placement of the Caerlaverock moat timbers.**

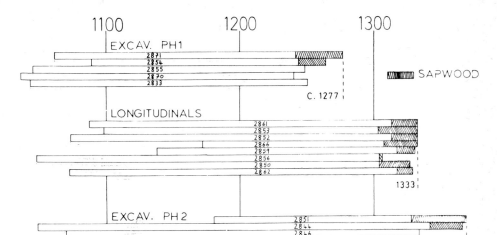

CAERLAVEROCK MOAT

battle. The castle was being refurbished. Whatever their primary function, these longitudinals were presumably re-used in conjunction with the Phase II bridge. The date of Phase II is 1371 or shortly after. While the sapwood of QUB 2851 was slightly damaged, the date is backed up by the estimated felling date of QUB 2842.

All the timbers discussed so far were characterised by narrow rings and long ring records. The Phase III bridge was of interest because of the completely different character of the timbers. While still of massive dimensions, up to 45 cm square cross-section, none exhibited a ring pattern longer than 59 years. It proved impossible to establish a date for these Phase III timbers. However, by their very fast-growing nature they would fit the description of regenerating timbers and since they must have been felled after 1371, it is reasonable to suggest that they, in keeping with other timbers from Castle of Park, Ulster, Dublin and elsewhere, represent further examples of early-fifteenth-century timbers which had regenerated after 1350 (see Chapter 11). While this Phase III dating cannot be proven, it seems to be backed up by the observation that the long drawbridge/abutment arrangement which replaced the bridges would fit with a mid-fifteenth-century extension of the gatehouse (Watson, 1922, 37).

*Dublin Medieval Excavations*

As outlined in Chapter 7, the presence of large quantities of oak timbers in late-twelfth- and thirteenth-century contexts within the medieval city of Dublin allowed the construction of a chronology spanning 855 to 1306 (Baillie, 1977c). Although this chronology spans the whole of the Viking period it was unfortunately of little use for the dating of structures from that period. In general terms the transition from Viking to Norman takes place shortly after 1170 and this is coincident with a change from ash to oak as the common heavy-timber element on the Dublin sites. This curious fact has still to be accounted for, but must reflect either preference on the part of the woodworkers or, more likely, lack of access for the inhabitants of Dublin to the oak forests which were later to be exploited by the Normans. So while research is currently under way to establish whether or not ash can be used successfully for dating purposes, the main effort in dating structures on the Dublin sites was restricted to oak and thus to the later levels on the sites. The examples cited below hopefully demonstrate how dendrochronological dates radically improve the understanding of archaeological contexts. The real reason for this improvement is the relationship between timber dates and human activity. For example, the interpretation of a coin find, while giving a useful spot date, tells us little about the context of loss. While we may know the date and place of manufacture very precisely, this precision rarely carries through to the archaeological findspot. With dated timbers on the other hand, it is possible to point to the act of felling, often in a specific year, which resulted in the presence of the timbers within a structure. Such intrinsic dating has advantages for tying together what might otherwise appear as rather disparate strands of evidence.

One of the first wooden structures to be studied dendrochronologically was Timber Framed Structure (TFS) 2 from Square 4, High Street. Structurally this feature took the form of a plank floor which had originally covered a pit. Pressure of the layers above had caused the oak planks, softened by the water-logged condition of the deposits, to settle into the cavity. Excavation revealed that a chute had protruded through the floor into the pit, which contained a quantity of fruit stones. This timber floor was analysed during the construction of the 452-year Dublin chronology, and was important because it was one of the few well stratified features which gave some clue about the date range of the floating chronology.

The archaeological dating evidence for TFS 2 consisted of a coin of 1199 at approximately the same stratigraphic level as the floor and a second coin of 1225 found 25 cm above.[5] The whole feature was sealed by a later wicker-wall and could not have been cut in. When the Dublin chronology was eventually tied down (Chapter 7) the felling dates of the various structures could be translated into real years. Fortunately, one timber from the floor, QUB 605, retained its total sapwood. This sample ended in Dublin Datum +72, i.e. 1201. When this is put together with the coin evidence we get a clear picture of a floor level, incorporating TFS 2, at a date very close to 1201 (paralleled by the coin of 1199) and the whole structure sealed by a layer containing a coin of 1225. It would be tempting to suggest that the lifetime of TFS 2 was something less than a quarter-century.

While the Dublin chronology was being constructed, two large timber-lined pits were sampled in order to establish their temporal relationship to the patently late-twelfth-/early-thirteenth-century TFS 2. These pits, TFS 4, High Street, and Pit 6/1, Winetavern Street, had each been only partially excavated. There were several reasons for this. First, each occurred in the baulks of their respective squares. Second, their positions, extending high up into the baulks, suggested that they might well be late, possibly even seventeenth-century in date. Furthermore, the fill removed from each in the partial excavation had yielded singularly little in the way of diagnostic dating evidence.

The timbers from Pit 6/1 yielded a chronology of 259 years, running out to the felling date of QUB 1124 at DD +104. The cross-correlation between the Pit 6/1 master and the available Dublin chronology was $t = 13.4$. So even initially, before the Dublin chronology was definitively tied down, the timbers from Pit 6/1 could be assigned a felling date 32 years later than those used in the construction of TFS 2 above. Clearly Pit 6/1 was thirteenth-century in date. The results for TFS 4, while highly comparable, show how inferior samples can significantly reduce the quality of the dating. The only sample from TFS 4 to retain its total sapwood, QUB 643, contained several exceptionally narrow bands in the outer portion of its ring record and more than 50 sapwood rings in the space of 2 centimetres. While the inner 70 years of its ring record agreed well with the other samples from the pit, the outer portion had an inherent error of (at a guess) ± 5 years. Thus the best estimate of the felling date of QUB 643 was in the range DD +97 ± 5. As corroborative evidence for this date another sample, QUB 639, exhibited a heartwood/sapwood transition at DD +69, and allowing for missing sapwood this indicated a date in the range DD +102 ± 9, highly consistent with QUB 643. So both these pits, on separate sites, had been constructed at essentially the same time, certainly within a decade in the early thirteenth century.

At least partially on the basis of their relative dates in the thirteenth century, now known to be 1233 and 1226 ± 5 respectively, the remainder of the fill was removed in each case. In Pit 6/1 an important hoard of 2,000 pewter tokens was discovered. These have been argued as belonging to the period close to 1279 (Dolley and Seaby, 1971, 448). In addition, in a layer which sealed the pit a bronze obole of Bordeau was found which dated to *c.* 1286-92. Pit 6/1 had been

**Figure 8.5: Schematic section Pit 6/1 Winetavern Street, Dublin.**

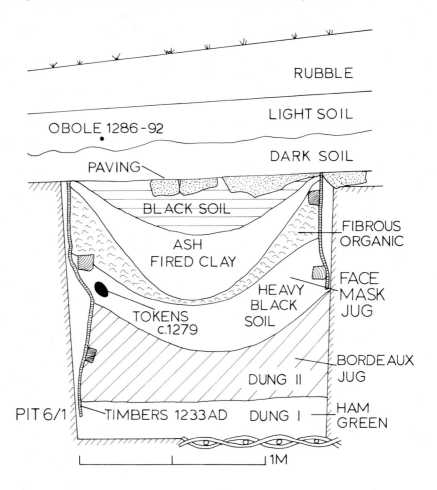

RUBBLE

LIGHT SOIL

OBOLE 1286-92

DARK SOIL

PAVING

BLACK SOIL

FIBROUS
ORGANIC

ASH
FIRED CLAY

HEAVY
BLACK
SOIL

FACE
MASK
JUG

TOKENS
c.1279

BORDEAUX
JUG

DUNG II

PIT 6/1   TIMBERS 1233 AD   DUNG I

HAM
GREEN

1M

Source: Courtesy B. ÓRíordáin.

built as a cess-pit. The fill consisted of three distinct layers, the bottom two of which were compressed organic detritus (see Figure 8.5). The third layer consisted of a fine black soil in which the tokens were deposited. In this same layer were fragments of a glass vessel and the neck of a thirteenth-century mask-jug.

Here then was the rare phenomenon of a pit with a construction date in the mid-1230s, an initial lifetime of around 40 years before the (presumed) deposition of the token hoard and a final sealing date suggestive of the late thirteenth century.

The real difference in this example is that the pit is not being dated, as is normally the case, by its fill. As can be seen at a glance, the finds in the pit would suggest a later-thirteenth-century date. It would be difficult to estimate the time between construction and the deposition of the more datable finds by any conventional means. Here we are seeing almost the complete history of a medieval cess-pit. In comparable fashion, at a point half-way up in the fill of Pit TFS 4 an almost complete mid-thirteenth-century English knight jug was discovered, again suggesting use over something like a quarter-century to half-fill one of these substantial pits (ÓRiordáin, 1973, 151). Obviously any question of the lifetime of the pits being extended by regular emptying could only be inferred by direct archaeological evidence for such practices.

As an example of the way in which dendrochronology tightens up slack archaeological thinking the case of the knight jug is salutary. It was originally thought by the author that this important jug had been found at the *base* of TFS 4. Had that been the case it would have moved the accepted chronology for such highly decorated wares back by a full quarter-century. Normally the accepted date is 1250 and after. Fortunately the excavator, when asked, supplied the information that the jug had in fact been found half-way up in the pit fill (ÓRiordáin, personal communication), thus saving considerable embarrassment.

Figure 8.6 shows to advantage the quality of the woodworking in such pits. This example, Pit 13/2, Winetavern Street, was so carefully made that intuitively it argued strongly for the use of fresh timbers (ÓRiordáin, 1971, 83). This is of course a major consideration in dendrochronological dating. 'Yes, you may be dating the timbers, but how do you know they were used immediately?' The normal answer with regard to seasoning is that in medieval and earlier times timber was used green. However, the question of re-use is rather more involved. If a structure of any kind were to be completely dismantled and reconstructed or used for some alternative purpose, then dendrochronology would date the original structure, not the reconstruction or later use. In the case of the pits, how can we decide whether or not the timbers have been used before? The answer has to lie in consistency of dating. For example, is it likely that the timbers in Pit 13/2 could have survived use and still been capable of inclusion in this pit exhibiting no sign of the earlier (?) use. Intuitively the answer is 'No', but can it be backed up?

To test this a selection of the planks and two of the uprights from Pit 13/2 were analysed. Several had at least some trace of sapwood and the relative datings are shown in Figure 8.7. All but one of the timbers appear to belong to the same period. The odd timber, QUB 706, retained its complete sapwood and was at least 20 years older than the remainder. This distribution of dates is certainly not random, as might have been expected in a structure composed of old or scrap wood. The bunching of the dates at around 1240 would suggest use at that time with the inclusion of a stray earlier plank.

**Figure 8.6: Axonometric drawing Pit 13/2 Winetavern Street, Dublin.**

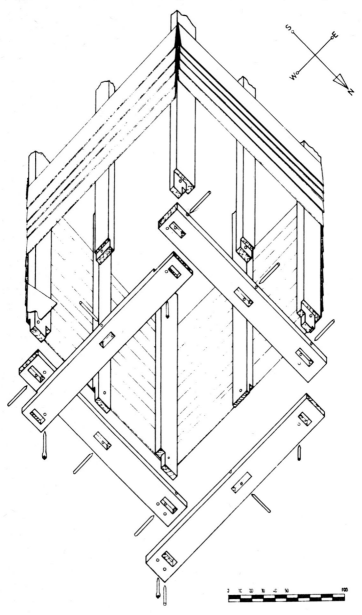

Source: ÓRiordáin (1971). Courtesy B. ÓRiordáin and *Medieval Archaeology*.

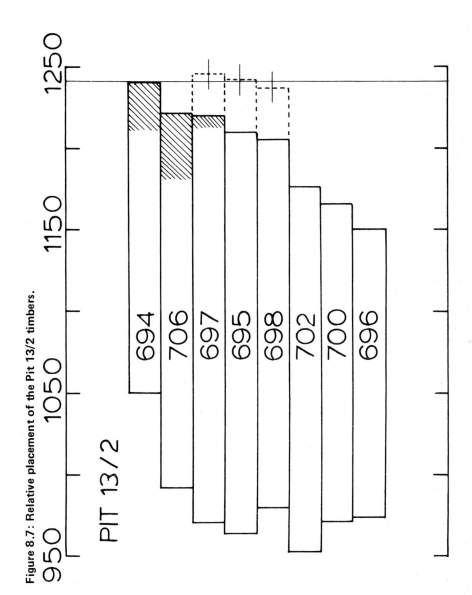

Figure 8.7: Relative placement of the Pit 13/2 timbers.

*Perth High Street*

Experience suggested that the best approach to the dating of timbers from a medieval excavation should normally be the construction of a 'site' chronology. All datings can then be relative from one timber or structure to another even if the chronology as a whole remains undated. In dealing with medieval excavations it should normally be possible, on the basis of associated finds, to assign an approximate age range to the site chronology even in the absence of a definitive cross-dating against an absolute master chronology.

At the time when timbers were becoming available from Perth, work was in progress on the construction of the 1,000-year Scottish tree-ring chronology. As indicated in Chapter 7, the timbers used in the construction of this chronology came from central and southern Scotland. The question of the overall area of applicability of the chronology, for dating purposes, awaited empirical testing. Thus the presence of oak timbers in Perth afforded an opportunity to observe whether cross-dating could be obtained between timbers from this vicinity and the available Scottish master chronology.

Initially it was hoped that sufficient long-lived timbers would be available from Perth to allow the construction of an independent Perth site chronology. This chronology could then be dated against the Scottish chronology, the procedure as used with Dublin, above. However, of the many Perth samples examined, only 15 contained what could be regarded as an adequate number of rings for possible dating. With the exception of a stray plank, QUB 3183, which contained 215 rings, the average length of the samples dated was only 120 rings. By comparison the timbers from Glasgow Cathedral, sampled for chronology-building purposes, averaged 200 rings. The relative shortness of the Perth timbers suggested that it might be difficult to build an independent Perth site chronology by the procedure of comparing the individual ring patterns. Instead each sample ring pattern was compared directly with the Scottish master chronology. Those samples which showed good visual and statistical agreement with the master chronology were then compared with each other at the specified positions of overlap. If these agreements were consistent, then the matching positions were accepted. In this way a total of 13 timbers were found to cross-agree, each matching with all of the others. Once all of the ring patterns had been dated they were run together to form a site master chronology. In theory, timbers found in Perth in the future should stand a better chance of matching with this site chronology than with the more remote Scottish chronology. The Perth chronology spans 256 years from 949 to 1204.

Now while the exercise of dating the timbers from Perth serves as a further example of the sort of procedure which can be used when dealing with material from a new context, one result in particular was of interest. As with the timbers from Glasgow Cathedral, the samples were treated randomly. Only find and context numbers were supplied with the samples. Table 8.1 lists four of the dates supplied for the site report. These four are abstracted from the list of 13 dated samples (Baillie, 1981a).

**Table 8.1: Perth Timbers from a Single House Structure**

| QUB Number | Perth Accession Number | Perth Feature Number | Number of Rings | 't' Value cf. Scottish Chronology | Date of Outermost Measurable Growth Ring | Date Quality |
|---|---|---|---|---|---|---|
| 3186 | A7334 | C3566 | 117 | 2.6 | 1150 | A |
| 3187 | A7336 | C3566 | 118 | 3.6 | 1150 | A |
| 3903 | A11568a | C5032 | 141 | 3.6 | 1150 | A |
| 3905 | A11568c | C5032 | 93 | 2.6 | 1150 | A |

Although it was not known at the time, when the site information was brought together in early 1980, it transpired that each of these samples was from the same structure. The message has to be that when timber samples are subjected to a standard analysis the results produced are repeatable at every level.

*Charcoal*

Although charcoal occurs on most archaeological sites, it is mainly in the form of small fragments. Occasionally, however, larger pieces such as the remains of completely charred posts or planks are encountered. In America this is a frequent occurrence and the methods employed in the preparation and measurement of charcoal samples have been repeatedly stated (e.g. Hall, 1946, 26). In the British Isles large charred sections of oak are not common but do occur. The main problem with oak charcoal is the tendency for it to split in three dimensions, radially, vertically and along the lines of vessels. This is exacerbated by the fact that in most cases the charcoal is found in a saturated condition. Once uncovered, it begins to dry out and this traumatic change can reduce seemingly substantial charred timbers to a heap of fragments. In the American south-west the charred timbers are almost invariably dry when encountered and hence they are relatively stable.

The problem when a substantial charred section is encountered is how to keep it together long enough for its ring pattern to be analysed. Plate 2(f) shows a completely charred plank section from High Street, Carrickfergus, Co.

Antrim. This sample was in a highly delicate state when excavated and all credit is due to the conservator who managed to extract it in a single piece. A simple expedient was to coat a strip of paper with 'Isopon' (a fast-drying car body-filler — and equivalent would do) and wrap the resulting bandage round the sample. Within minutes the coating was hard and rigid. The sample is still intact nearly six years later. The ring pattern of this sample, QUB 4224, was measured on the freshly broken transverse surface without preparation of any kind. The resulting 120-year pattern, made up of three overlapping measurements, included total sapwood, still identifiable by its hollow vessels. The ring pattern cross-matched with the Belfast chronology with its outer ring equivalent of 1551 ($t$ = 4.9). The date was of interest since the plank occurred as part of a charcoal spread which was cut by the foundation trench of a tower house, believed to have been erected between 1560 and 1567 (Delaney, 1976, 5). Charcoal can therefore provide perfectly adequate dates if encountered in conditions suitable for measurement. Hindsight suggests that photographs of the samples *in situ* might well allow the piecing together of ring patterns which otherwise might become too disarticulated for reconstruction. It is worth remembering that there may well be important sites where a little extra care with charred timbers could allow precise dating as an alternative to radiocarbon.

Most of the examples cited above were undertaken for two reasons. The first and most important was the acquisition of ring patterns for chronology building. The second was the clarification of dating problems. Now that the chronologies are in existence it should be possible to direct the method towards the resolution of specific dating problems. There is an unfortunate tendency for people in the British Isles, both dendrochronologists and others, to regard the method as such a novelty that to date anything becomes an end in itself. This is regardless of whether the sample or structure is of any importance, whether or not its date is already known, whether or not the date has any value historically or archaeologically. An analogy would be the early days of the cinema when the moving picture was worth watching for its own sake, regardless of content. Once the basic chronologies are available, and in essence that situation has already been reached, it will be essential for the method to mature; the novelty can only last so long. In order to prove its worth, problem areas of chronology will have to be isolated and research programmes formulated which will answer some of the really important questions in history and archaeology.

This aspect of dendrochronology finds an analogue in the radiocarbon calibration saga. Since the early 1970s and the furore over the initial calibration attempts, everyone has been waiting for the definitive high- precision calibration. The irony is of course that when such a calibration does become available in a few years' time, certainly by the mid-1980s, it will serve only to calibrate other high-precision dates in any meaningful way. Routine radiocarbon determinations

**Figure 8.8: Schematic elevation of Trier Cathedral with tree-ring dates tracing the building development.**

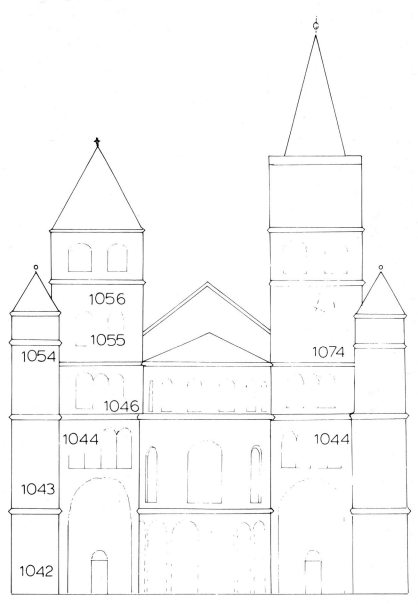

Source: After Hollstein (1979, 162).

can be calibrated, but the uncertainties associated with the dates themselves will render interpretation almost impossible. Let us hope that the users of radiocarbon dates realise what *they* are up against (see also Chapter 12).

So, as a final example, and to show what can hopefully be achieved in a formulated approach to a dating problem, it is worth looking at Trier Cathedral. In Germany chronologies covering the last millennium have been available since the 1960s. By dating selected timbers from Trier, Hollstein was able to follow the progress with the construction of the cathedral in the felling dates of the building timbers (Hollstein, 1979, 162). This exercise does not require inter-pretation: Figure 8.8 speaks for itself. There seems no obvious reason why similar results should not be entirely possible for at least some of the more prestigious buildings and archaeological sites in the British Isles and elsewhere.

## Notes

1. The historical information is not absolute in any sense; it is just that a date in the 1640s would be incompatible historically.

2. There seems no reason to doubt the 1570 date since it is carved into the overall design on the beam along with Bathe's name.

3. Obviously in such conditions the sapwood has to be ignored and only the heartwood/ sapwood transition considered.

4. Obviously $1576 \pm 9$ is still within 2 standard deviations of 1590 and so is still accept-able. However, we are trying to improve the felling estimate and clearly it is not working in this case.

5. The author is indebted to Mr Brendan ÓRíordáin for making such information freely available.

# Chapter 9

# Early Medieval Chronology

The next period back in time, the first millennium AD, includes the aptly named Dark Ages — dark if in no other sense than dendrochronologically. In Britain and Ireland this general period includes the Saxon period in England and the Early Christian period in Ireland. From a dendrochronological viewpoint the problems change in the early medieval period. The archaeological chronology of the first millennium is still weakly developed. It would be reasonable to suggest that in the case of most sites and excavations of this period it was not usually possible to assign an accurate date on the basis of archaeological finds alone. Thus when attempting to build a tree-ring chronology in the first millennium it was not immediately possible to draw on well dated sources of timber. On the contrary, it was frequently up to the dendrochronologist to establish the dating of particular sites. In this vein radiocarbon was widely used to establish the broad date range of sections of chronology. Radiocarbon dates had to be treated with caution, however, since the whole question of applicability of the American calibration to the Old World had still to be resolved (see Chapter 12).[1]

While the availability of published German chronologies for the second millennium had aided English workers by tying down their medieval chronologies, the lack of published German chronologies for the first millennium severely limited progress with Saxon dendrochronology in England. Fletcher has pointed out that most of the Dark Age material studied in England has tended to be fortuitous (Fletcher, 1977). This merely reflects the fact that no one had set out to construct an English Dark Age chronology and material had not been searched for in any systematic way. Thus in England the construction of a chronology for this period had to await the completion of an absolute chronology in some adjacent area. At Belfast, the knowledge that it would probably not be possible to cross-date directly with German chronologies, even when they were available (see Chapters 6 and 7), prompted effort to be applied to the construction of an independent Irish chronology for the early medieval period.

Interestingly, as early as 1957 Schove was attempting to tie down sections of English Saxon chronology. This involved attempts to recognise hypothetical signatures in the tree-ring patterns. These signatures were predicted on the basis of expected responses of oak trees to certain aspects of recorded weather information: for example a significantly narrow ring was searched for to coincide with an extensive summer drought recorded for the year 764 (Schove and Lowther, 1957, 81).

Such attempts were doomed to failure for the simple reason that there could be no guarantee of the response of an oak to any particular aspect of climate. Moreover, it would seem a dangerous course to tie down any section of ring pattern on the basis of a single narrow ring. Remember that throughout the work described in these chapters total ring patterns have been used as a basis for cross-dating. In addition, Schove restricted the period of time within which he looked for this hypothetical cross-dating. Variously on the basis of archaeological information, coin evidence and radiocarbon dating (both raw and calibrated) he altered the general period of possible dating (Schove, 1974, 1979). In retrospect these dating attempts showed a failure to understand that, in the final analysis, the dating of an ancient tree-ring pattern can only be correctly specified by its cross-agreement with an already dated chronology. Only by adhering rigidly to this principle can the absolute precision of the method be ensured and, equally important, its essential independence.

Up to the time of publication of Hollstein's chronologies (Hollstein, 1979) no published tree-ring sequence was in existence which spanned the first millennium in Europe. Published chronologies were restricted to the last 1,000 years in round figures (see Table 9.1). On this basis Schove cannot be blamed for trying to tie down Dark Age ring patterns by means other than dendrochronology. It was simply an ill-advised exercise. The price paid was to be consistently wrong.

**Table 9.1: Maximum Extent of Published Oak Chronologies pre-1979**

|  | Date Published | Year AD | Location |
|---|---|---|---|
| Hollstein | 1965 | 822 | Germany west of Rhine |
| Huber and Giertz | 1969 | 832 | Germany central |
| Delorme | 1972 | 1000 | Germany Rhineland |
| Baillie | 1977c | 855 | Dublin |
| Baillie | 1977a | 946 | South-east Scotland |
| Baillie | 1977b | 1001 | Northern Ireland |
| Fletcher | 1977 | 782 | South-east England |

The above sets a backdrop against which the construction of early medieval chronologies for the British Isles during much of the 1970s must be viewed. No seriously dated tree-ring chronologies earlier than 780 were available in England and no first-millennium material whatsoever in Scotland or Wales. For Germany it was known that Hollstein's chronology extended back to around 700 BC, but this was not available. (Unpublished sections of chronology covering the period 207 to 746, 441 to 1460 and 578 to 1234 for south-central Germany, northern Germany and Scandinavia were communicated personally to the author in the late 1970s by Bernd Becker, Dieter Eckstein and Thomas Bartholin respectively.) Unfortunately their usefulness for dating sections of Irish oak chronology was seriously questionable on the basis of experience in the medieval period. The best that could be expected was a stepwise agreement, Ireland to England to Germany, and in the absence of suitable English chronologies the construction of an Irish first-millennium chronology had to be undertaken in isolation.

### The Belfast First-millennium Chronology

It was clear that available timbers of the medieval period throughout the British Isles tended to have ring patterns which extended back to the ninth or tenth centuries (see Table 9.1 and Chapter 11). It was and is extremely difficult to acquire timbers of the tenth to twelfth centuries whose ring patterns might extend back beyond the ninth century. As late as 1979 only one timber, a sample from Tudor Street, London, positively extended back further. This timber spanned the years 682 to 918 (Hillam, personal communication) and was to be of fundamental importance, as will be seen below.

   In Ireland, there are no standing buildings with timbers which might belong to the tenth or eleventh centuries. Moreover, it is impossible to specify archaeo-logical sites of these centuries which might yield timbers.[2] So the logical backward extension of chronologies again had to give way to an approach similar to that adopted by Douglass, i.e. the construction of an earlier floating chronology forward in time (see Chapter 1). Hopefully such a chronology would eventually be extended sufficiently to cross-match with either the Dublin 855 or Belfast 919 absolute chronologies. The starting point for the construction of a floating first-millennium chronology was a group of timbers from an Early Christian period site at Teeshan, Co. Antrim. This site, a crannog or artificial island, was destroyed in the late 1960s and the opportunity was taken to collect a large number of random timbers. For those not familiar with crannogs some descriptive information is included in Appendix 1. At the time of destruction stray archaeo-logical finds suggested that Teeshan had been in use in the mid-first millennium. In fact the group of timbers, when analysed, yielded a floating chronology of 494 years. A number of radiocarbon determinations confirmed that this chronology covered broadly the first five centuries AD. Since this chronology was floating

**Figure 9.1: (a) First-millennium progress by 1973. (b) First-millennium progress by 1975. (c) First-millennium progress by early 1979.**

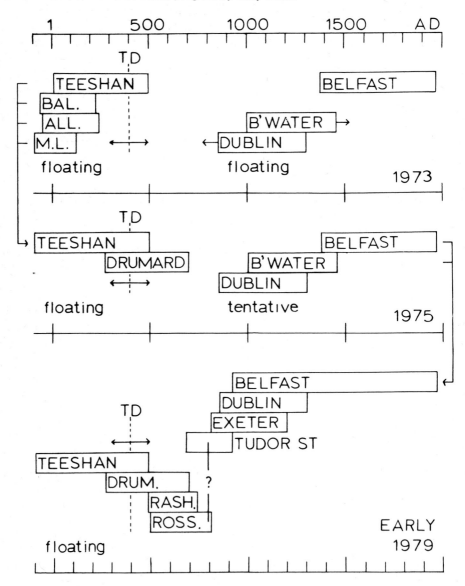

and since it was hoped to extend it in two directions, an arbitrary scale was necessary. In practice, one year was chosen as Teeshan Datum (TD) and the chronology was labelled relative to this year. Thus the Teeshan site chronology spanned TD-390 to TD +103. A word of caution is necessary at this point with regard to the dating. While placed roughly in time, it was recognised that the chronology could easily move a century or more in either direction, as neither the archaeology nor the radiocarbon dates were in any sense reliable as indicators of true age.

Two sources of material were to play a significant role in the development of the Teeshan chronology in the first instance. These included timbers from natural contexts and a second crannog. A typical group of bog oaks came to light in the demolition of a saddler's cottage at Balloo, Co. Down. These timbers had been dug out of a local bog in the late eighteenth or early nineteenth centuries and used as roof beams. Several of the timbers cross-matched to form a 296-year chronology (see also Chapter 10). Since the oaks could in theory have been of any age, samples were submitted for radiocarbon dating. It transpired that the chronology belonged to the first millennium. Bluntly, this had not been expected, since the majority of bog oaks are prehistoric, so no attempt had been made to compare the Teeshan and Balloo chronologies. When compared, the Balloo chronology showed strong cross-agreement with Teeshan, with its outer year equivalent to TD-166 ($t$ = 5.6). Similar findings were made with another group of sub-fossil oaks from Allistragh, Co. Armagh, which spanned TD-439 to TD -141 ($t$ = 5.1).

Not all sites provided such neat solutions as Balloo and Allistragh. A large assemblage of sub-fossil and archaeological timbers from a crannog and its environs at Mill Lough, Co. Fermanagh (also known as Killyfoal Lough) produced dates from 4000 BC to 1500 AD for various site chronologies. This diversity made analysis of the assemblage very difficult and a clear picture was obtained only after considerable effort by Jennifer Hillam. Two sections of Mill Lough chronology were of particular importance for quite different reasons. One section of 222 years cross-matched with Teeshan, extending the chronology back in time to TD -490 ($t$ = 4.9). The other was important not because it extended the chronology but because it was composed of worked, hence archae-ological, timbers. This section of Mill Lough chronology ended at TD +75 ± 9, suggesting archaeological activity on both sites at broadly the same time. Overall, by 1974 the floating chronology covered 594 years.

While this progress was encouraging, there was no doubt that several centuries separated the 594-year chronology from the oldest end of the, by then, tentatively dated Belfast and Dublin chronologies (see Figure 9.1(a)). With such a large gap to fill it would have been preferable if some sub-fossil material had dated to the later first millennium rather than the earlier half. Three groups of bog oaks had matched with the Teeshan chronology. In the general 6,000-year chronology building project (see Chapter 10), numerous miscellaneous timbers were dated by radiocarbon. Up to 1978 none had produced dates in the second half of the first millennium. It was beginning to look as if, between say 600 and 1000 AD, conditions were not conducive to the preservation of oaks. If this were indeed the case, the onus would be firmly on archaeological contexts to provide the necessary timbers for chronology building.

The state of first-millennium dendrochronology at Belfast remained static during 1974 and 1975 awaiting the acquisition of some suitable timbers. Ironically, the key to a major step forward with the Teeshan chronology had been residing in the reserve wood-store since 1973. In that year a routine drainage operation involved the deepening of a stream bed at Drumard, Co. Londonderry. In the course of this operation a series of heavy black oak timbers was thrown up by the mechanical excavator. These timbers included the flume of a horizontal mill. The flume, which consisted of a large hollowed-out oak log, was observed by some schoolchildren who reported it to their local schoolmaster. He visited the site and identified it as an ancient mill. Although some stray wood samples were taken for dendrochronological analysis, the full significance of the find was not immediately apparent. Basically, in 1973 people were not conditioned to thinking automatically of these structures as early.

The horizontal mill is a beautifully simple piece of technology. In essence it is a powered quern. The water-wheel, driven by a jet from the flume, lies in a horizontal plane and drives the top stone of the mill directly via a vertical shaft (see Figure 9.2). A more detailed description is given in Appendix 2. It is not known for certain where the idea developed, but it is likely that it originated, like so many other inventions, in China. It was certainly known in Europe by the beginning of the Christian era and tradition ascribes its introduction into Ireland to the third or fourth century (McAdam, 1856). The complicating factor lay in the known occurrence of these mills, still in operation in the Islands of Scotland and in western Ireland as late as the nineteenth century. Thus archaeologists were conditioned to think of horizontal mills as a continuum of which most would be late rather than early. Those which had turned up in Ireland during the last century had almost without exception failed to yield any object which could be firmly dated. In England, on the other hand, the horizontal mill at Tamworth (see below) had been dated by radiocarbon to the eighth century (Rahtz and Sheridan, 1971) and had been hailed as important at least partly because of its early date. The Irish mills lacked any prestige connection and no individual mill could be assumed to be early.

The outcome of this attitude was a failure to appreciate the possibility of an early date for the Drumard timbers. The few timbers sampled in 1973 were compared only with the Belfast 1380 to 1970 chronology. (A rather naïve

**Figure 9.2: Suggested reconstruction of the Drumard horizontal mill. Lettered timbers represent surviving elements.**

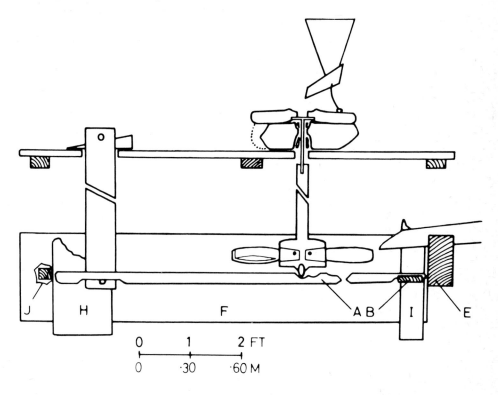

expectation existed that, because the timbers showed signs of burning, Drumard might have been destroyed in the 1641 Rebellion (see Chapter 5).) Fortunately the timbers did *not* cross-date with the Belfast 1380 chronology.

It was not until 1975 that the realisation dawned — horizontal mills must fall within the last two millennia, they cannot be prehistoric, therefore ring patterns derived from them must be of some use for chronology building. What was needed was a proper representative selection of timbers from the mill, sufficient to allow the construction of a site chronology. Thus in 1975 the schoolmaster, Mr George McIlroy, was revisited and samples removed from a selection of timbers which he had stored against the day when a reconstruction might be possible. These, together with further samples which had survived *in situ* in the stream bed, yielded a 428-year Drumard chronology. This chronology cross-dated directly with the Teeshan chronology with an overlap of 227 years ($t = 6.9$). This extended the Teeshan complex forward in time from TD +103 to TD +304, a two-century extension at a single stroke (Baillie, 1975). Figure 9.1(b) shows the state of development of the chronologies at Belfast in 1975.

### Placement of the Floating Chronology in Time

By 1975 a floating chronology of 795 years was in existence covering approximately the first eight centuries AD. It is perhaps worth labouring this point about the dating of the chronology. Although it contained timbers from archaeological structures, no close archaeological dating could be assigned to the ring patterns. Since no relevant chronologies from either England or Germany were available at the time, placement had to depend on radiocarbon dating, but there were problems in attempting to tie down a chronology using this method.

The 795-year chronology had associated with it a number of radiocarbon dates. These were conventional radiocarbon dates using the 5,568-year half-life. To fix this chronology in real time these dates had to be converted to historical years, i.e. they had to be calibrated. The only available calibration curves for the first millennium were based principally on the results of Suess (1970). It must be remembered that it was not proven that the American calibration produced dates compatible with Old World historical chronology. Conventional wisdom suggested that calibration of radiocarbon dates in the first millennium moved the dates towards the present, i.e. the raw radiocarbon dates were too old. However, a line of argument championed by Alcock (1977) and Campbell *et al.* (1979) suggested that calibration of radiocarbon dates for 'known age' samples from Dundurn and Cadbury in fact made the calibrated dates less compatible with historical chronology rather than more compatible as would be expected.[3] If this suggestion were correct, then rather than calibrate Old World radiocarbon dates of the first millennium (i.e. make then younger), it might be more appropriate to convert them to the new half-life (5,730 compared with 5,568), which would have the effect of making the raw dates older by a factor of 1.03.

This discussion left us with a 795-year floating chronology which was dated by radiocarbon but which could move either forward or back in time depending upon how one tried to interpret the radiocarbon dates. From the point of view of the dendrochronologist who wished to know the width of the gap between the floating and fixed chronologies, this situation was less than satisfactory. It became obvious that the chronology could not be completed until such time as a definitive tree-ring match was obtained between the absolute and floating chronologies. This cross-match, when obtained, would override all radiocarbon considerations. For practical purposes it was assumed that the 795-year chronology covered the approximate period 50 BC to 750 AD. It was recognised that this dating could be wrong by up to 100 years either way. The search was on for suitable timbers to resolve the dating.

*Possible Sources of Timber*

In order to complete the Belfast chronology for the last two millennia, timbers of the second half of the first millennium were necessary. It seemed reasonable to suggest that the likely sources of such timbers would be crannogs and, possibly, horizontal mills. The other ubiquitous first-millennium site type, the rath or ring fort, consistently fails to turn up substantial timbers. The only example to do so in recent times, Lissue, Co. Antrim (Bersu, 1947) is unfortunately represented in the Ulster Museum collections by no more than a handful of wood fragments, of which only two hold out hope of dating under any circumstances.

Archaeologically the consensus opinion on crannogs would suggest broad usage in the Early Christian and medieval periods, the medieval occupation being borne out by the results given in Chapter 7. The postulated broad Early Christian usage of these sites was based on the results of a relatively few excavations and was conditioned by finds from possibly long periods of occupation. In the hope that crannogs would supply suitable gap-bridging timbers, a number of these sites were visited during the mid-1970s. Wherever possible, exposed structural timbers were sampled. It was assumed that if there had genuinely been activity on these sites throughout the later first millennium, then this random sampling would yield a wide spectrum of dates. In practice, the felling dates produced from random timbers from identifiable Ulster crannogs were anything but random. Although late dates were being sought, all the timbers sampled belonged to the period between TD +75 and TD +150 (see Table 9.2). This clustering of dates from crannogs suggests that the use of oak timbers was largely primary on the sites. Later occupation in the first millennium, if it existed, did not give rise to oak remains. On the basis of these findings crannogs were abandoned after 1978 as useful sources for chronology building (Baillie, 1979a).

**Table 9.2: Relative Dates for Definite Crannog Sites in the North of Ireland**

| Site | County | Felling Date |
|------|--------|-------------|
| Teeshan | Antrim | TD +103 |
| Tamin | Antrim | TD +140 ± 9 |
| Midges Island | Antrim | TD +92 ± 9 |
| Island McHugh | Tyrone | TD +144 ± 9 |
| Mill Lough | Fermanagh | TD +75 ± 9 |
| " " | " | TD +120 ± 9 |
| Ross Lough | Fermanagh | TD +92 ± 9 |

Broadly speaking, this left horizontal mills as the only likely source of late-first-millennium timbers. After all, Drumard represented by far the furthest extension of the floating chronology forward in time. Now, just at the time that crannogs were falling from favour, a second horizontal mill site was discovered and regrettably destroyed, at Rasharkin, Co. Antrim. The flume of this mill, found during excavations for a sewage works, had excited local interest as a 'giant's coffin' — a not altogether ridiculous suggestion given that this large hollowed oak trunk had a close-fitting longitudinal lid held in place with wooden pegs (see Plate 8). Samples from the salvaged mill timbers, including the flume which exhibited total sapwood, were cross-dated and a Rasharkin master produced. This site master cross-dated with the Teeshan/Drumard chronology ($t$ = 7.4) and extended forward in time to TD +344. So the timbers used in the two mills had been felled only 40 years apart in time.

Subsequent to the Rasharkin find, it was discovered that the flume of a mill which had turned up, at Maghnavery near Markethill, Co. Armagh, in 1962 was still in existence. This mill flume, quite remarkably similar to the example from Rasharkin, had survived due to the offices of a colourful local woman, Mrs Qua, who at the time of the original finding had cautioned the present owners: 'Never get rid of that boat.' Thus the flume had been preserved, known locally as *Granny Qua's Boat*. This flume yielded a 200-year ring pattern which cross-dated with the Drumard/Rasharkin chronology with its outer year at TD +300 ($t$ = 4.8). Allowing for missing sapwood, a reasonable estimate of the date of construction at Maghnavery would be TD +332 ± 9, highly consistent with the Rasharkin date.

The dating of these examples reinforced the notion that horizontal mills could be a useful source of timbers for the later first millennium. There were, however, two rather worrying factors. The first was the obvious bunching of these mills, since the only three to be precisely dated had all been constructed within 40 years. The second was in the nature of these sites. They are only ever discovered accidentally, and it is therefore impossible to go out looking for horizontal mills or to predict where they will be found. Their chance discovery has to be awaited. In particular the bunching was worrying, as it suggested a possible phase of mill building around TD +300 to TD +340. If this were true it might reduce the chances of finding later examples.

However, in 1979 attention was drawn to the flume of a horizontal mill in the collection of the National Museum of Ireland, originally from Rossorry, near Enniskillen, Co. Fermanagh. This flume, which had at some stage been misinterpreted as a dug-out boat, was figured by McGrail (1978, 125) in his corpus on British Isles log boats together with a correct identification. A series of cores taken from this flume yielded a 306-year ring pattern which cross-dated with the three existing mills, Drumard, Rasharkin and Maghnavery with $t$ = 3.0, 3.4 and

**Plate 8: Flume of the Early Christian horizontal mill at Rasharkin, Co. Antrim. The tree from which it was cut had last grown in AD 822.**

Source: Crown copyright, photo by Dr A. Hamlin.

3.6 respectively. Importantly, this sample extended the floating Teeshan/Drumard chronology forward to TD +416. (Since the outer year of several of the cores from the flume was consistent, it can be suggested that it represents the heartwood/sapwood transition. Thus allowing for missing sapwood the felling date would most probably lie in the range TD +448 ± 9.) This ring pattern extended the floating chronology to a total length of 907 years.

Here then was a major step forward in the extension of the floating chronology. If the placement in time of the Teeshan/Drumard chronology was even remotely correct, this extension forward to TD +416 suggested that Rossorry should overlap with the dated Tudor Street timber (start year 682, above). Unfortunately no significant cross-agreement could be found between these two chronologies. It must be remembered of course that Rossorry and Tudor Street were represented by the ring patterns of single trees which grew some 500 km apart. Should agreement be expected?

Up to this point all of the first-millennium chronology construction had been confined to the north of Ireland. One reason for this was the relatively unknown nature of cross-matching within Ireland as a whole. However, by the beginning of 1979 the study on modern trees from the whole of Ireland was available (see Chapter 4). The results of that study lent hope that cross-dating could be obtained to at least some extent throughout Ireland. This was to prove important, since the extension of the floating chronology to TD +416 had effectively exhausted the supply of later-first-millennium timbers within the north of Ireland. Figure 9.1(c) shows the early 1979 situation for the first millennium.

## Southern Irish Horizontal Mills and the Dating of the Floating Chronology

The finding that all four dated horizontal mills from the north of Ireland had been constructed within a period of 150 years (TD +304 to TD +448 ± 9) raised one interesting point: if this clustering of mills in the second half of the first millennium was general to Ireland, then all that was necessary to construct a parallel Early Christian chronology for the southern half of Ireland was to sample timbers from a range of these structures within that area. Moreover, if a southern chronology was constructed, the possibility existed that it might consolidate the 'gap' between the floating chronology and the Dublin chronology (starts AD 855) or alternatively cross-date directly with London, Tudor Street.

While it was impossible to search out new horizontal mills, for reasons given above, one obvious source of mill timbers was the published examples (Lucas, 1953, 1955; Fahy, 1956). It was possible that at some of the sites timbers had been left *in situ*. This was a reasonable hope given the remote locations of some of the published mills. However, in 1979 the key to a more systematic sampling exercise became available in the guise of a list of horizontal mills, drawn up by the National Museum of Ireland in 1970. This contained the findspots of a number of these structures which had turned up, mostly in the southern half of Ireland during extensive drainage operations in the 1950s and 1960s. By coincidence in 1979 three new examples became available for study in Co. Cork. Here then was the opportunity to test two hypotheses. First, could the timbers from horizontal mills from diverse parts of Ireland be dated against the Belfast chronologies, and, second, would the horizontal mills throughout Ireland conform to the tight dating of the northern examples? Equally important was the possibility of obtaining timbers which would allow consolidation of the floating 907-year chronology (Baillie, 1981b).

The results outlined below are concerned with the dating of timbers from nine horizontal mill sites distributed in the southern half of Ireland. No attempt is made to analyse or record the structure of these sites, for that remains the province of the respective archaeologists who recorded details during their original examinations. The dating of the timbers and the construction of a southern chronology are our principal concern. For the sake of brevity only the experiences with those sites relevant to the outlining of the chronology are recounted.

The first site investigated was Ballykilleen, Co. Offaly (National Museum No. 10). This mill had been excavated by Lucas in 1953 (Lucas, 1955). Would the timbers still be in existence in 1979, 26 years later? It is probably worth recounting the experience of relocating the Ballykilleen timbers as a lesson to others (not just for Ballykilleen but for any previously excavated site). It was possible with an adequate grid reference to navigate to within one field of the site. Casual approach to the nearest farmhouse yielded a guide to the exact findspot in a deep drainage ditch. No timbers were visible and the dimensions of the ditch suggested that the total site had been removed. It would have been reasonable to depart at that point and scratch Ballykilleen from the list. However, the guide pointed out that the findspot was actually on his neighbour's land and suggested that the owner of the site be contacted. Arriving at the owner's farmhouse, his three sons admitted no knowledge of the mill nor of any ancient timbers. This seemed conclusive until the farmer arrived in person. He not only remembered the mill but also the original excavation. In addition he was able to relocate two of the massive squared beams from the mill —overgrown in a ditch 100 m east of the original site. The timbers were in excellent condition despite being out of the ground for a quarter-century. This demonstrated the tenuous nature of even local knowledge.

Of the two timbers, QUB 3529 and 3530 with 153 and 172 rings, only the latter showed a heartwood/sapwood boundary. When compared with the Teeshan complex both cross-dated giving $t = 5.9$ and 6.3 with their outer rings equivalent to TD +90 and TD +126 respectively. Allowing for missing sapwood on QUB 3530, the felling date should lie in the range TD +158 ± 9. Clearly Ballykilleen had been constructed considerably earlier (about 150 years) than the northern mills. This was encouraging in that it smeared the date range for mills and lent hope that an even wider range of dates might be forthcoming. In particular it was a useful demonstration that timbers from well outside the original northern area of study could be cross-dated with the Teeshan/Drumard chronology.

The early dating of Ballykilleen was confirmed by timbers from a mill at Little Island, Co. Cork. It is possible that this mill is the same one recorded in the National Museum list as having been uncovered at Little Island townland in 1803 (National Museum No. 55). This extremely large example turned up in the summer of 1979 and appears to have been powered by tidal changes rather than the more usual stream, river or spring. This Little Island site yielded a chronology of 370 years which again cross-dated with Teeshan/Drumard with its outer felling year being TD +152 ($t$ = 6.0). Within the limits of the method these two mills could have been of identical date.

Fortunately, from the chronology-building point of view, these were the only two early examples encountered. The others conformed more closely to the northern distribution. The most important southern mill initially was that at Brabstown, Co. Kilkenny (National Museum No. 27). This mill, originally discovered in 1964, was preserved as two side beams and a cross-piece in the bed of a stream at Brabstown. Again, as at Ballykilleen, it could not have been rediscovered without local assistance. Samples from the side beams indicated that they were the two halves of a large oak, split and partially squared. This effectively reduced the number of trees represented to two. The two samples cross-dated with each other and with the Rossorry ring pattern. The side-beam pattern QUB 3691 contained 282 rings without sapwood and gave $t$ = 5.6, 3.1 and 3.1 when compared with Rossorry, Teeshan/Drumard and Maghnavery respectively. In each case its outer ring was equivalent to TD +403. Allowing for missing sapwood, a suggested felling date would be in the range TD +435 ± 9, highly consistent with Rossorry.

There were two principal reasons for the importance of Brabstown. First, it gave a fixed section of chronology from the far south which was to act as a useful reference for tying down most of the other mills. Second, its ring pattern ran out to TD +403 *without* sapwood. Since this was close to the furthest extension of the floating complex, at TD +416, if samples from Brabstown could be obtained with sapwood they would presumably give a crucial extension of the complex to something like TD +435 ± 9.

By late 1979 the situation was becoming clear. None of the other southern mills showed any sign of extending the chronology.[4] In addition, none of those dating with the floating chronology showed any definitive sign of cross-dating with Tudor Street. The situation is summarised in Figure 9.3(a). In order to break out of this impasse a further collection trip was organised in early December 1979. There were two objectives. The first was to revisit Brabstown and attempt to find sapwood on some of the timbers, and second was to try for samples from the few mills known in the extreme south-east, in Waterford and Wexford.

At Brabstown two fresh timbers were discovered. A squared beam QUB 3865 lay at the end of one of the longitudinals and must originally have been an upright.

**Figure 9.3: (a) First-millennium progress by late 1979. (b) Consolidation of the first millennium via Ballydowane and London Tudor Street (Hillam).**

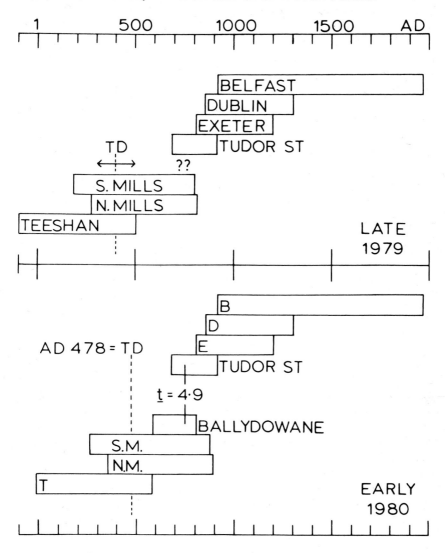

This sample exhibited definite traces of sapwood. In addition a riven oak upright was uncovered behind the same longitudinal. This sample, QUB 3864, retained its total sapwood. On the basis that most of the mills appeared to be single-period sites, it was reasonable to suppose that these samples would extend to the felling date of QUB 3691. The remainder of the trip produced samples from Horeswood, Co. Wexford (NMI 35) (as yet undated but with less than 90 rings) and Ballydowane West, Co. Waterford (see below) (NMI 42).

The return trip to the laboratory was in the sure and certain knowledge that we had cracked the problem. Since it could be assumed that the floating chronology ran out to at least the ninth century, the extension offered by the sapwood opened up the possibility of a direct link with the Dublin 855 chronology. The samples QUB 3864 and 3865 were prepared and measured wet in order to hasten the extension. The two new samples cross-dated to give a total ring pattern of 212 years. Imagine the consternation when the mean of these two cross-matched, not at the end of the floating chronology as expected, but with the outer year at TD +282 ($t$ = 7.1). As a result of this blow the Christmas break of 1979 was spent in a state of gloom based on the quite definite feeling that the 800 to 900 AD gap was not intended to be bridged. Thus it was not until well into January 1980 that the single timber from Ballydowane was finally measured. Curiously, very little hope was held out for dating this timber on account of its isolated position on the extreme south coast.

Ballydowane had originally been discovered in 1970. It survives as a single morticed beam QUB 3682 which had been deposited in a hedge some distance from the original findspot (local information again). Its ring pattern of 221 years cross-dated with Brabstown QUB 3691 ($t$ = 5.7) and with Rossorry ($t$ = 4.2) with its outer heartwood ring at TD +331. Allowing for missing sapwood, the felling date should have been in the range TD +363 ± 9. So again there was to be no extension of the chronology: Ballydowane fell firmly in the main period of mill building (see Figure 9.4). The real importance of this timber was not realised until its ring pattern was compared with Hillam's Tudor Street 682 to 918 pattern. The Ballydowane ring pattern showed a good visual and statistical match against Tudor Street with an overlap of 128 years ($t$ = 4.9). If this dating were correct it specified TD +331 on the Irish floating complex as AD 809, making Teeshan Datum the year 478. This key timber fulfilled the same role for Irish dendrochronology as HH 39 had done for Douglass in 1929. So ironically the December trip had supplied the necessary link, but not in the way we had expected. Table 9.3 lists the Irish horizontal mills currently dated. A further five sites have as yet failed to date definitively — almost certainly due to their short ring records. Given the geographical spread of the mill sites (see Figure 9.4), it would seem unlikely that clustering of the dates in the second half of the first millennium is a biased picture. Moreover, within the 300-year range it is interesting to note

**Figure 9.4: Distribution of the dated Irish horizontal mills.**

DATED HORIZONTAL MILLS

AD 630-930

that 70 per cent lie in the century AD 770 to 870. On that basis the clustering of the three first northern examples was a genuine reflection of the overall distribution.

**Table 9.3: Relative and Absolute Dates for Irish Horizontal Mills**

| Site | County | Date of Outer Ring (relative scale) | AD Date |
|------|--------|-------------------------------------|---------|
| Little Island | Cork | TD + 152 | 630 |
| Ballykilleen | Offaly | TD + 158 ± 9 | 636 ± 9 |
| Morrett | Laoighis | TD + 292 | 770 |
| Drumard | Derry | TD + 304 | 782 |
| Ballyrafton | Kilkenny | TD + 316 ± 9 | 794 ± 9 |
| Maghnavery | Armagh | TD + 332 ± 9 | 810 ± 9 |
| Ballygeardra | Kilkenny | TD + 333 ± 9 | 811 ± 9 |
| Rasharkin | Antrim | TD + 344 | 822 |
| Ballydowane West | Waterford | TD + 363 ± 9 | 841 ± 9 |
| Keelaraheen | Cork | TD + 365 | 843 |
| Farranmareen | Cork | TD + 395 ± 9 | 873 ± 9 |
| Brabstown | Kilkenny | TD + 435 ± 9 | 913 ± 9 |
| Rossorry | Fermanagh | TD + 448 ± 9 | 926 ± 9 |

## Confirmation of the Early Medieval Chronology

The good cross-dating between Ballydowane and Tudor Street had to be seen for what it was; essentially the matching of two individual ring patterns from a considerable distance apart. An obvious course was to look for confirmation to this match. It was already known that no good cross-dating obtained between Tudor Street and the other Irish timbers. Obviously that was not the place to look. There was, however, another possible direction. It had already been noted that a good cross-agreement existed between Tudor Street and a floating chronology, Ref 8, produced by Fletcher (1977). This chronology was made up from the ring patterns of timbers from Old Windsor and Porchester and covered 322 years.[5] Hillam had also noted this Ref 8 to Tudor Street match, which was extremely good but unfortunately rather short, 56 years ($t = 3.8$). On its own this could not be considered definitive. However, if it was correct it suggested 737 as the outer year of Ref 8. It is important that the logic of the following argument should be understood.

If the Irish early medieval complex TD -490 to TD +416 spanned 13 BC to 894 AD, as indicated by the Ballydowane/Tudor Street cross-match, and if Ref 8 matched Tudor Street with its outer year equivalent to 737, could any consistent match be found between Ref 8 and *any* of the Irish material? It must be remembered that no doubt whatever attached to the internal (relative) matching of the material from the 13 Irish horizontal mills. If, therefore, any of the constituent Irish timbers cross-matched significantly with Ref 8 at a self-consistent date, the case would be proven. In fact three Irish ring patterns, those from Ballydowane, Maghnavery and Ballygeardra, showed good cross-dating with Ref 8 at their specified dates with $t = 3.5, 3.4$ and $3.3$ respectively.

So here was a completed circle of agreement (not to be confused with a circular argument) between Tudor Street, Ref 8 and Ireland. Since Ref 8 was now fixed in time, as a further check on this dating, could any agreement be found with the German chronologies of Becker and Eckstein cited above? Again it must be remembered that agreement was being sought at a specific position, i.e. with Ref 8 ending in 737. In fact Ref 8 gives a correlation of $t = 3.7$ against Becker's Donau 5 chronology at exactly 737, this match also being visually acceptable. Here then is a date for Ref 8, specified on the basis of cross-matches with English and Irish chronologies, being corroborated against an independent German chronology. Figure 9.5 shows these various correlations plus further corroboration supplied by a chronology from Mersea (Hillam, personal communication) and another from Tamworth (see below).

In restrospect it is easy to see how the evidence was accumulating for the whole of the early medieval period. Although many of the cross-matches were known about, their full relevance did not become apparent until Ballydowane specified the exact dates for comparison. The consistent replication of this key match makes the dating of the early medieval complex absolutely certain.

## Tamworth Horizontal Mill

With the dating of what was now a complex of Dark Age material, 13 BC to 894 AD for Belfast, a parallel southern Irish chronology spanning AD 260 to 881 and Ref 8 covering AD 416 to 737, an obvious course was to attempt the dating of some known timber-rich site or sites. One site in particular stood out as relevant. This was the horizontal mill at Tamworth excavated by Rahtz in 1971 (Rahtz and Sheridan, 1971). Since there existed a clear grouping of Irish horizontal mill dates (Table 9.3), it was considered relevant to attempt the dating of this mill for comparison. Although radiocarbon dates had been obtained in 1971, these dates could not be compared directly with the more precise tree-ring dates. So in obtaining samples from Tamworth there were two objectives. The first was to attempt a dating and the second was to relate any date obtained to the context of the other horizontal mills in the British Isles, namely those in Ireland.

**Figure 9.5: Cross-correlations linking the independent British Isles chronology complex to Germany.**

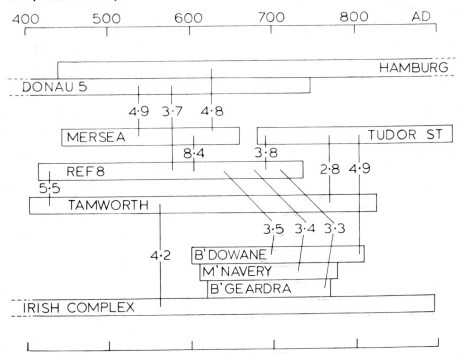

Wood samples from the Tamworth mill were kindly supplied by Claire Tarjan, together with some further samples excavated in 1978 by Bob Meeson at a site believed by him to be related to the leat of the same mill. The results were as follows. All the timbers from both sites cross-dated to give a 422-year Tamworth chronology. This chronology cross-dated directly with the consolidated Belfast chronology with its outer year corresponding to AD 825 ($t = 4.2$). The only three samples with heartwood/sapwood boundaries ended in the years 825, 824 and 820, and thus, allowing for missing sapwood, it can be suggested that the mill complex was built in the range 855 ± 9 (using the Belfast sapwood estimate). Meeson's site and the mill itself were indistinguishable in date.

Clearly the Tamworth mill was constructed within the main period of horizontal mill building in Ireland. Thus it may not be the exceptional structure it seemed at first sight. A date c. 855 clearly rules out construction under Offa (died 796), though it fits in well with the evidence of the charters which reached their peak in the 840s and the 850s when more were signed at Tamworth than at any other town in England.

In terms of chronology, Tamworth extends precisely dated English tree-rings back to AD 404. As a still further check on the replicated cross-agreements between Ireland and England, Ref 8 also cross-matches with Tamworth with its outer year equivalent to AD 737 ($t = 5.5$) (Baillie, 1980, 63).

## Notes

1. Conventional radiocarbon dates are calculated from carbon sample activity on the theoretical assumption that activity decreases continuously with age. Observation has shown that the true relationship between sample activity and age must be deduced from an empirically derived calibration curve. During the 1970s the only available calibration for the first millennium was based on American analyses of known age wood samples.

2. These may seem like rather sweeping statements, but they are unfortunately true.

3. However see Chapter 12 — the actual calibration relationship may be much worse for the interpretation of first-millennium dates than was ever feared.

4. A separate study initiated by Lucas has shown broadly the same date range using radiocarbon determinations (Otlet and Walker, 1979).

5. Ref 8 had been dated to the ninth century by radiocarbon, though this could not be relied upon, as it placed a rather naïve faith in ordinary radiocarbon dates (see Chapter 12).

# Chapter 10

# Prehistoric Dendrochronology

Prehistoric is used here in a rather loose sense to incorporate the study of both sub-fossil and prehistoric archaeological material. The whole of the tree-ring programme at Belfast was initiated as a response to the availability of large numbers of sub-fossil oaks — bog oaks — in the north of Ireland in the late 1960s. The initial step was to assess the possibility of cross-matching tree-ring patterns in a British Isles context. After all, up to 1970 no information was available to show whether or not dendrochronology would work in this region. Once it had been established that cross-dating could be obtained between oaks (see Chapters 4 to 6), the second stage was to assess the possibility of building a long sub-fossil chronology. The approach to this was to take random samples from different sources and subject these to radiocarbon analysis. This very quickly showed that there existed a fairly continuous spectrum of bog oaks throughout the last 6,000 to 7,000 years. The samples dated were selected at random, but had they shown a strong tendency to cluster the project might well have been abandoned.

Figure 10.1(a) shows the situation as it existed in 1972. It was clear that in terms of oak availability most periods were represented. The dates spread back for a considerable portion of the total period over which oak is known in Ireland. The rational limit for oak on most pollen diagrams lies between 7500 and 9000 bp (Smith and Pilcher, 1973, 906). The theoretical maximum length of any oak chronology in Ireland would be some 9,000 years. In practice, since the earliest deposits will be the least accessible, the likely maximum length expected, even from the outset, was something of the order of six to eight millennia. One factor which should be removed at this point is consideration of pine for dendro-chronology in Ireland. In the north of Ireland, if the number of bog oaks could be numbered in the thousands, by comparison the number of bog pines could be counted in the tens of thousands. There are, however, two significant reasons why they were not considered for chronology building. The first is the plain fact that there are no pines available in the north of Ireland between 2000 bc (radio-carbon years) and 1700 AD. Pine (Pinus sylvestris) disappears from the pollen record at 2000 bc, presumably as a result of some climatic change, though it is not impossible that some other factor may have been responsible (Smith and

Pilcher, 1973, 911). It would therefore be impossible to build a continuous pine chronology. A random sample of pines dated in the same exercise confirmed that the dates cluster at the older end of the age spectrum. The second disadvantage would apply even to the construction of a floating pine chronology. This relates to the ability of pines to exhibit more than one ring per year, and worse still to miss rings at least locally around their circumference. These factors make it necessary for multiple measurements to be made on each sample to establish a usable mean ring pattern for each tree. Pilcher, in constructing floating pine chronologies within the north of Ireland, found it possible to obtain adequate results if three radii per tree were measured and meaned. This threefold increase in effort was found to be the limiting factor. Pilcher did, however, succeed in producing floating chronologies up to 600 years in length (personal communication). These pine chronologies have not yet been exploited to any purpose, although with the upsurge of interest in dendroclimatology, especially through the medium of densitometry, these chronologies could still have important applications (see Chapter 12).

Having settled on oak as the species for study, how was the chronology building initiated? The procedures had to be somewhat different from those associated with the study of historic building timbers. In the latter case the age of the timbers was known to within decades, or at worst one or two centuries. With bog oak the spread of uncertainty was of the order of thousands of years. It has to be understood that the bog oaks were seldom sampled *in situ*. In most cases the oaks came to light as a result of some commercial activity, typically dredging, land reclamation, peat cutting and motorway construction. Normally the oaks would become available for sampling as unstratified heaps containing anything up to 50 trees.

In only a few cases was it possible to sample stratified oaks where they occurred in the sides of ditches. It was to become apparent, from results produced over a period of years, that stratification meant very little, for example two oaks lying at the same 'level' in a bog section could be of quite different ages, depending on the bog stratigraphy. In extreme cases vertically stratified oaks, lying in physical contact, had died fully 500 years apart. This failure of stratigraphy to be more than the most general guide to relative age was partly due to the ability of the trees to slice into the bog when they fell and to settle until their downward progress was arrested. In this respect it is not surprising to find an oak trunk resting on top of another which had died half a millennium before.

Probably the most serious factor concerning the lack of useful stratigraphy is the shrinking of most bogs due to drainage. Over the last century a considerable amount of drying out has been induced by drainage. The result is that oaks, dotted through the layers of a raised bog (like fruit through a cake), are moved together as the peat shrinks around them. Local farmers in the fenlands to the

Figure 10.1: (a) Progress with the Belfast 6,000-year chronology 1971. (b) Progress with the Belfast 6,000-year chronology 1974. (c) Progress with the Belfast 6,000-year chronology 1977. (d) Current status of the Belfast 7,500-year chronology. (e) Current status with the German long chronology.

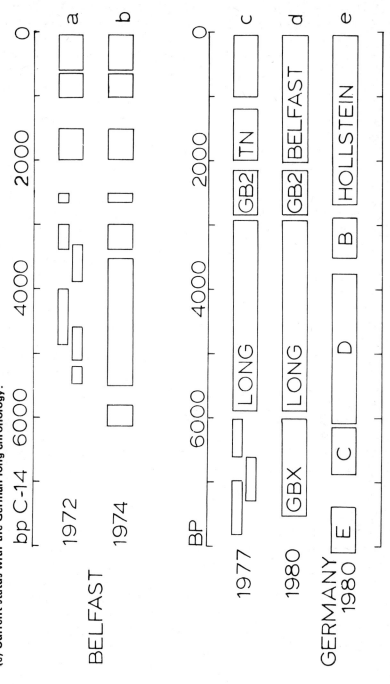

south of Lough Neagh describe the bog oaks as 'floating', i.e. each year they drag the oaks off the surface of a ploughed field, and the next year a new crop has 'floated' to the surface. This is in fact due to the reclaimed bog continuing to shrink around the oak trunks, making them appear to rise.

Why are the oaks preserved? There are two aspects to this question: the first is why were oaks growing on peat – a rare phenomenon today; the second is how did they survive to become bog oaks? To deal with the second point first: it is impossible to estimate the proportion of the original population of oaks growing on the bog surface which became buried. It is possible that the proportion, despite the large number of bog oaks, was quite small. From the ring patterns of the oaks and from their surviving dimensions it can be surmised that they were in general tall narrow trees with narrow rings (averaging 1 mm or less). This may indicate close growing conditions, but another explanation may be that the narrow rings were caused by poor levels of nutrient in the peat.

It appears that the trees grew on living bog. When a tree died it might stand and rot *in situ*, only the stump being buried (see Plate 9(a)). If, on the other hand, it blew down or was dragged down by one of its neighbours falling (it seems reasonable to assume that oaks growing on peat would not be as securely rooted as those on mineral soil, though of course there is no proof that this is the case) then the peat could grow around and bury it. The condition of many of the trunks shows that the tree fell into the peat leaving its top surface exposed. Exposure was sufficiently long for the bark and sapwood to rot off the upper surface while the buried surface has both intact. On some occasions exposure was sufficient for the complete upper half of the trunk to decay, leaving a flat surface. So the general picture must have been falling, rotting and eventual burial, the speed depending on the rate of peat growth.

Once buried in the anaerobic conditions of a bog, preservation would be assured. Over a period of time, iron present in the ground-water would complex with the tannins in the wood to produce the characteristic black colouration. The degree of blackening depends heavily on the amount of iron present in the ground-water. Thus colour has little to do with age in samples more than a few hundred years old. For all practical purposes, colour of bog oak is no indication of age. This is of course very unfortunate.

The other factor relating to bog oaks concerns the conditions which encouraged oaks to grow on open bog, sometimes rooted on peat several metres in depth. The two most likely causes are periodic drying out of the bog surface, and conversely, periodic flooding of the bog. If the surface dried out, sufficient rotting might take place to release nutrients and support tree growth. More attractive, however, is the notion of periodic flooding. Most raised bogs are low-lying and frequently have rivers running close to or through them. Flooding would have the effect of depositing minerals and rendering the otherwise inhospitable bog surface suitable for colonisation by oaks. What we do know is that bogs were capable of supporting individual trees for up to 450 years. They could support trees continuously for periods up to the order of thousands of years. Equally well there are glimpses of information showing that at some periods oak growth was interrupted simultaneously on several bogs (see Chapter 11) (Baillie, 1979, 26).

**Plate 9: (a) Typical *in situ* bog oak at Ballymacombs More, Co. Londonderry (second millennium BC).**

**Plate 9: (b) Typical group of unstratified bog oaks from the fenlands to the south of Lough Neagh (fourth millennium BC).**

## Sampling and Chronology Construction

From the late 1960s onwards, every opportunity was taken to collect groups of bog oaks whenever they occurred. Only the most general information was available on likely age. For example, if it could be seen that oaks were stratified below pines then it could be surmised that the oaks were at least 4,000 years old. Normally even with samples still *in situ* the only hope of estimating age (short of radiocarbon dating) lay in the analysis of pollen samples from beneath the trunk. By comparing the proportions of species in the spot sample with known, dated pollen profiles, it was possible on occasion to suggest a very rough age range. Plate 9(a) shows a typical *in situ* sample. More often the trees occurred as unstratified heaps where they had been dragged to the edges of fields or work-sites (see Plate 9(b)). On these occasions no dating information was available.

Sampling was carried out mostly using a chainsaw. Two slices were taken, wherever possible, from each tree: one slice about 3 or 4 cm thick for dendrochronological study and a reserve slice between 15 and 30 cm thick as a source of both samples for radiocarbon calibration (see Chapter 12) or other research as yet unspecified. It was realised early on that this material represented a unique data source and, once dated, might be of use for a number of lines of enquiry. Unfortunately it could not be known until after the samples had been processed which trees were really important. Often by the time this was known the original material had been dumped, re-buried or burnt. It was necessary therefore to treat all samples equally in the field and this had the unfortunate effect that important samples were represented by a reserve no bigger than that kept for those which turned out to be least useful. It was obviously important to number both slices identically and hopefully permanently so that the samples could be traced back to their sources. This may seem trivial, but with a reserve of some thousands of samples any unnumbered slice of bog oak could be of any age from any site and might as well be discarded.

Having acquired the raw material, these slices formed the building blocks of the various floating chronologies. The process of chronology building did not always move forward smoothly. As with all research, many facets of this study progressed simultaneously and the results were naturally unpredictable, tending to occur in quantum jumps, i.e. no progress despite continuous effort, then a sudden leap forward. Because of this it would be unrealistic to attempt a step by step description of the research. Rather it is proposed to describe only the major steps in the process of chronology building. This will be illustrated with the details of the construction of a site chronology followed by the cross-dating of two site chronologies and finally the merging of a number of site chronologies to yield the overall situation.

First let us consider the nature of the problem to be solved. We had a number of sites and from each site a number of individual trees. In general terms these trees could be of any age within the last 7,000 years. In the case of two individual trees it was not known whether they were the same age or different ages. If they were of different ages, it was not known in which order they might come in time, i.e. which was older. Confronted with this apparently random picture, how was chronology building initiated? In practice, an assumption had to be made. Rather than treat all bog oaks as a group and search for cross-matches, it was assumed that *within the random grouping of trees from a single site it was more likely to find some which were contemporaneous.* Had this not been the case, the possibility of ever building a long sub-fossil chronology would have been negligible. The exercise reduced therefore to the construction of site chronologies.

In order to test the above assumption large numbers of ring patterns had to be measured and compared. It has to be pointed out that almost all the original work on sub-fossil oaks at Belfast, the measurement and matching, was carried out by Jennifer Hillam over the period 1970 to 1976. It is thanks to her painstaking and methodical work that the chronologies outlined below exist. The method was to choose a site and measure two trees. Their ring patterns were

**Figure 10.2: Individual ring patterns from Lisnisk with the resultant site master.**

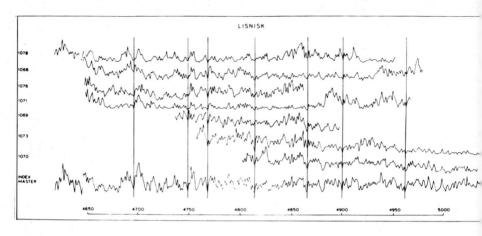

Source: Pilcher *et al.* (1977).

compared visually and by computer correlation. On average each would be around 200 years in length. If these two did not cross-date, they were accepted as two units and a third pattern from the same site could be compared with both. If neither matched, there were three units with which to compare the fourth pattern, and so on. Eventually two ring patterns would be found to cross-match acceptably (sometimes as many as 20 trees were compared before two would match). Once a match was obtained a mean ring pattern was prepared and each of the unmatched patterns re-run against the mean.

Eventually by processing all the ring patterns from a site a simplified picture would emerge. For example, there might be two submasters and half a dozen unmatched individuals. If this situation could be reached, two things had happened. One, the number of variables in the puzzle had been reduced and, two, the quality of the pieces, i.e. the submasters, had been improved. A submaster should always contain more matching signal than an individual ring pattern. For the moment the unmatched individuals could be set aside, though they would have to be studied further at a later date because of several possibilities. Consider one unmatched ring pattern:

(a)    it could be the same date as one of the submasters from the same site but would not match either because it was wrongly measured or it was a poor ring pattern (for some reason);

(b)    it could be of a different date from the submasters and would eventually match with some other site chronology;

(c)   it could fall across a gap or be otherwise useful;
(d)   it would never match with anything and would join the 'guesstimated' 10 to 20 per cent of non-matching material.

As for the submasters, because they were made up from ring patterns which cross-matched it was possible to have more confidence in them than in individuals. Therefore radiocarbon dates would be obtained for the submasters to place them roughly in time. At first progress was very rapid: after all, there was a 7,000-year empty span into which each new section of chronology could be placed. However, as the number of chronology sections increased there would come a point when there were more chronologies than available time. At that point, if not before, some of the various submasters must cross-date. This is in fact what happened. The established procedure became the construction of site submasters, comparison of these with all already available submasters and, if no cross-dating could be obtained, recourse to radiocarbon dating. So the chronology-building process had moved away from individual ring patterns to a situation where site submasters were being treated as the site individuals had been originally. Figure 10.1(b) shows how the situation had developed by 1974. Time was filling up to the point where, on the basis of the radiocarbon dates, many of the extended chronology sections should have been cross-matching.

Here an additional factor had to be taken into account. This was the incompatibility between radiocarbon years and real (tree-ring) years. Suess's calibration results had shown (Suess, 1970) that radiocarbon dates in the BC era were too young by up to half a millennium. So in Figures 10.1(a) and 10.1(b) the time scale is artificially compressed. There was in reality more time into which the submasters could fit than was suggested by the radiocarbon dates. It was important to take account of this, since otherwise there were chronology sections which should have overlapped but which patently did not. Expanding the time-scale eased what would otherwise have suggested a serious flaw in the method.

The logical outcome of this chronology-building process was the eventual filling of the last seven millennia. Unfortunately filling seven millennia is not the same thing as completing a 7,000-year chronology. Reference to Figure 10.1(d) shows the 1980 situation. The whole chronology is complete with the exception of three 'gaps'. The real problems begin at this stage. The law of diminishing returns has set in and dictates that most fresh material processed will simply duplicate the existing chronology sections. Material to bridge gaps is difficult to obtain for the simple reason that it is *specific* material. Since the whole programme was based on random sampling, any search for specific trees (specific in the sense of the time they must span) goes against the whole sampling process. The nature of the problem has changed. It is hard to find trees which grew from 1000 to 800 BC — where are they likely to be? It is impossible to say. This aspect is covered in more detail in

Chapter 11. Having set the scene, some specific aspects of the construction of the prehistoric chronologies are outlined below (see Pilcher *et al.*, 1977).

## Site Chronologies and the Long Chronology

A fairly typical group of sub-fossil oaks was found by chance at Lisnisk, Co. Down. The trunks had been removed from the surface of a field to facilitate ploughing and had been drawn into a heap. It was clear from observation of the surrounding area that some depth of peat had previously been removed from the farmland. This suggested that the level from which the trees were coming was of considerable age, having originally been deeply buried. Fourteen samples were collected and processed in the manner outlined above. Out of the 14 samples 2 were found to have come from the same tree (their ring patterns being identical) which must have been broken in two at some stage and hence sampled twice. Of the 13 trees, 7 cross-dated to form a submaster of 421 years. Another group of 3 trees cross-matched to give a shorter submaster of 230 years. The two submasters did not cross-date. Of the remaining 3 samples 2 were narrow-ringed and about a century old (and hence borderline in usefulness), the remaining sample of 262 years remaining undated. So Lisnisk represents a good example of the simplification of a group of 13 trees down to two replicated chronologies and one individual ring pattern. The individual ring patterns in the seven-tree chronology are shown in Figure 10.2. Table 10.1 lists some of the correlation values between pairs of trees as an indication of the levels of agreement found.

**Table 10.1: Correlation Values between Individual Ring Patterns from Lisnisk, Co. Down.**

| | | | |
|---|---|---|---|
| QUB 1066 — QUB | 1073 | $t =$ | 6.3 |
| 1066 — | 1076 | | 8.7 |
| 1066 — | 1071 | | 15.8 |
| 1066 — | 1078 | | 5.8 |
| 1069 — | 1073 | | 4.9 |
| 1069 — | 1070 | | 4.2 |
| 1073 — | 1070 | | 6.8 |

Source: Pilcher *et al.*, 1977.

Radiocarbon analysis of a sample from one of the Lisnisk trees indicated that the longer site chronology lay in the early fourth millennium BC (calibrated radiocarbon age). A chronology building block had thus been created and placed approximately in time.

The second site to be considered, Balloo, Co. Down, demonstrates the diverse sources from which samples were collected. The bog oaks from Balloo had served as roof timbers in a domestic building since the late eighteenth or early nineteenth century. This reflects the scarcity of fresh timbers available at that time, as outlined in Chapter 5. A total of 30 timbers were collected. After removal of duplicates and unsuitable ring patterns these yielded two site chronologies. Radiocarbon analysis indicated that one lay in the first millennium AD (see Chapter 9). The other, of 5 trees, was shown to be of similar date to the longer Lisnisk chronology. Computer comparison of the Balloo and Lisnisk chronologies gave a correlation value of $t = 7.7$ with an overlap of 202 years between the chronologies. Thus two chronology building blocks had yielded a continuous replicated chronology to over half a millennium in length. It must be stressed that the use of radiocarbon was to provide no more than a rough guide to age. The tree-ring match was the sole basis for specifying the exact relative positions of the two site chronologies.

As the chronologies developed it became necessary, as with the medieval and Dark Age floating chronologies in Chapters 5 to 9, to specify a time-scale. In the case of the long chronology covering three millennia a set of numbers was assigned to the chronology running opposite to the BC dates, hopefully to avoid confusion. Such a relative scale is essential to allow reference to the chronology, to allow for its extension at either end and for computer comparison.

The result of the successive cross-matching of some 15 site submasters was the construction of one chronology covering 2,990 years. This chronology is believed to cover almost exactly the period 900 to 3900 BC. Some of the constituent site chronologies are shown in Figure 10.3. In addition to this long chronology there are two other important floating complexes. These are Garry Bog 2 (GB2), which spans the approximate period 100 to 800 BC, and Garry Bog X (GBX) which spans a 1,500-year period around 4000 to 5500 BC. Various lines of research are aimed at joining these chronologies definitively.

*Equivalent German Chronologies*

The bog oaks used for chronology building in Ireland had, almost without exception, been preserved in peat. In Germany there are also very large numbers of sub-fossil oaks; these occur, however, in the gravel terraces associated with ancient river courses. Whereas the Irish oaks are mostly preserved where they grew, the German oaks are mostly the results of erosion phases in the last 9,000 years. In all other respects the approach to the building of chronologies has been very similar. Three principal workers have been involved in different parts of Germany. Becker has concentrated on the south exploiting the Danube, the Upper

**Figure 10.3: The basic site chronologies which formed the building blocks of the Belfast 'long' chronology, spanning approximately 900 to 3900 BC.**

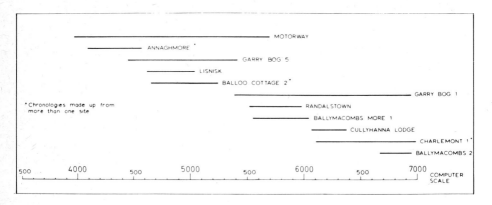

Note: a. Chronologies made up from more than one site.
Source: Pilcher *et al.* (1977).

Rhine and the Main. Delorme has worked in the north-east on the Weser and Schmidt has developed the more westerly Rhineland (Becker and Delorme, 1978, 59). Their results have broadly been produced in collaboration and for the sake of simplicity the German work can be considered as a unit.

The oaks have become available from both commercial gravel exploitation and dredging. They are individually of very comparable length to the Irish material, though there is a tendency for them to be wider-ringed. This may be due to their having grown on mineral soils rather than peat. It is possible for Becker to duplicate most of Hollstein's chronology which ran back to 700 BC. In fact the sub-fossil chronology for the period 207 to 745 AD appears to have been instrumental in correcting an error in Hollstein's chronology in the fourth century AD (see Chapter 12).

As outlined from the Belfast work, the progress in filling up time with approximately dated floating chronologies also proceeded rapidly in Germany. One disadvantage of working with oaks from river growth does appear, however. Whereas oaks can grow for long periods on a single bog, the forces at work in the evolution of river systems are much more dramatic. This has caused the overall chronology picture in Germany to be rather more fragmented. This probably reflects the short sharp episodic nature of river gravel deposition in comparison with the inexorably slow growth of a raised bog. Figure 10.1(e) compared the 1978 overall German chronology with that for Belfast in 1980 (Becker and Delorme, 1978, 61).

One obvious question relates to the possibility of using German chronologies to link the Belfast floating chronologies or vice versa. While this may be possible, it is by no means certain. However, what is becoming clear is that the high-precision radiocarbon work taking place on both the German and Irish oak chronologies will allow the dates of these chronologies to be specified to within very narrow limits in the relatively near future (see Chapter 12).

### Dendrochronology and Prehistoric Archaeology

There is no doubt that dendrochronology is destined to have a significant impact on archaeological dating in the first and second millennia AD. Principally this is because the results are fully compatible with historical documentation. However, in the era before written history it is questionable whether dendrochronology as a direct dating method is all that important. It is more likely that its principal contribution to prehistoric dating will be through the medium of radiocarbon calibration — that is, to the better understanding of the one universally applicable dating method.

The sentiment expressed here may seem pessimistic, and it certainly needs elaboration. Perhaps it would be better stated that dendrochronology is not going to make any significant impact on prehistoric archaeology in the British Isles. The real difference between the Irish and German chronologies lies in their potential archaeological usefulness. Broadly speaking, in Europe there are considerable numbers of prehistoric sites which produce adequate timbers. This is not the case in the British Isles, for not only are timber-rich prehistoric sites scarce, but those which do exist tend to be somewhat irrelevant in terms of precise dating. Becker and Delorme (1978, 61) can illustrate a 513-year chronology comprising two sub-fossil chronologies and ring patterns from *ten* Swiss and West German Late Bronze Age sites. It is doubtful if sufficient timber-bearing sites exist in the British Isles to allow the production of a similar diagram by the year 2000 (if ever).

Certainly in Ireland it is clear that in the last decade no significant pre-Iron Age site has produced useful timbers. The one site worth dating where even a vestige of timber remained was Ballynagilly, Co. Tyrone (Apsimon, 1976). This Neolithic long-house yielded charred oak planks. Unfortunately this was before the tree-ring work got seriously off the ground and no attempt was made to save the charcoal for dendrochronology. However, assuming that it would have been technically possible to resurrect the ring patterns from these timbers and date them with the floating chronologies, the question of what a precise date would mean for a Neolithic house in Ireland remains. It would be very nice, but it would be irrelevant; being the only precisely dated site its usefulness would be strictly limited in an environment where all chronology is based on radiocarbon.

Consider the one prehistoric site so far tied down against the long (2,990-year) chronology. This was Cullyhanna Hunting Lodge, Co. Armagh (Hillam, 1976). The site comprised a timber palisade and a hut circle and must have been a Bronze Age seasonal hunting camp beside a lake. The 313-year Cullyhanna chronology cross-dates with the 2,990-year chronology at a point which must be very close to the year 1500 BC (real date). The irony is that this site which could archaeologically have been of any date from Bronze Age to medieval, yielded only four flints apart from the structures themselves. A date to the second millennium BC would have been more than adequate for such an enigmatic structure. Now what is being said is not that such sites should not be dated by dendrochronology — why not use the method since it is there? — but rather do not expect such dating exercises radically to change or refine our picture of prehistory.

However, in case the picture is becoming too gloomy let us look briefly at one aspect of prehistory which can be elucidated by dendrochronology. This relates to internal or relative dating. It will be possible from time to time on the few sites which do turn up timbers to make some estimate of length of occupation on the basis of internal dating. This would involve looking for evidence of re-use or rebuilding which might well show up in the timbering. Unfortunately, at Cullyhanna all that could be said was that the hut and surrounding palisade were the product of a single felling (Hillam, 1976). One British Isles site where relative dating has been used to advantage, but where again absolute dating would be largely irrelevant, is in the Somerset Levels (Morgan, Coles and Orme, 1978). In such a context useful information on methods of exploitation and internal development can be gained.

*Chapter* 11

# Gaps in Tree-ring Chronologies

During the chronology-building process there has always been a tendency for archaeologists and others to ask, 'Where are you back to now?', meaning how far did the chronology extend back in time compared with the last time they had asked. This question was couched in the expectation that if the chronology extended back to 1380 at Christmas then obviously by the following Easter an extension would have been produced and another by Hallowe'en. This line of thought was built on the natural assumption that there exists a continuum of material for chronology building. In point of fact the chronology extended to 1380 by 1972. When asked the question in 1973, 1974, 1975 or 1976 the answer remained the same — 1380. Immediately the fourteenth century was bridged, the chronology ran back to 1001 and this was extended almost immediately to 919. The chronology for Dublin, also specified in 1977, ran back to 855. That situation remained static during 1978 and 1979. It was not until 1980 that the crucial link was forged back to 13 BC. Doubtless it will be some time before the next extension, back to around 900 BC, will be completed. Progress over the last decade clearly shows evidence for quantum jumps rather than relentless progress.

These observations immediately open up the notion that the material for chronology building and its availability are not continuous. Empirical observation suggests that for some periods there are very few trees available. The failure for so long to acquire ring patterns to bridge the fourteenth century led to common usage of the term 'fourteenth-century gap'. Undoubtedly the factors affecting timber availability will be complex. There may just be sufficient evidence to suggest that consistent gaps relate to phenomena which might better be specified as depletion/regeneration periods.

## The Seventeenth Century

While it would not be correct to say that the seventeenth century constituted a gap in tree-ring terms, it does show clearly what the dendrochronologist might expect to observe from a period of depletion/regeneration. During the construction of the Belfast chronology, described in Chapters 4 to 7, the longest-lived modern oak available had started life around 1649. It is necessary here to define to some extent the terms to be used in this chapter.

The dendrochronologist is normally concerned with observation of the felling years of trees. It is possible, as demonstrated in Chapter 8, to specify the final year of growth precisely in many instances. On the other hand, during the construction of chronologies the attempt is being made to measure as far back as possible, i.e. to observe the rings closest to the inside or pith of the tree. Unfortunately there will normally be a difference between the innermost ring of a tree and the year of planting. This difference may amount to quite a number of years depending upon how far along the trunk of the tree (vertically) the ring pattern is observed. So it is possible to talk loosely about the innermost ring of a tree-ring pattern as the onset of growth, although it is not by any means the same thing. What the dendrochronologist is observing as the innermost ring, even assuming the pith is being observed, is a date before which growth must have been initiated; a *terminus ante quem* for the tree. In the following discussion terms such as 'started life' or 'onset of growth' will be used. In each case what is being referred to is the innermost observable growth ring on a particular sample.

To return to the seventeenth century, the earliest observed growth ring on any modern oak in the north of Ireland was for the year 1649. The furthest extension forward in time of the outermost growth ring of any oak from a historic building in the same area was 1716. This provided a minimal overlap, the confirmation of which was covered in Chapter 6. Historic information, outlined in Chapters 4 and 5, gave a good basis for understanding the nature of this minimal overlap. The forests available in 1600 were systematically exploited for industrial and building purposes throughout the seventeenth century (McCracken, 1971). So there was a depleted stock of oaks by the early years of the eighteenth century. This, combined with the enclosure of large amounts of land which effectively removed most remaining oaks from the building stock, accounts for the failure to find oak timbers in any building after 1716. Coupled with this was the tendency of landowners in the eighteenth and nineteenth centuries to plant oaks specifically to replace the depleted stocks. This represents the 're generation' which in other periods may have taken place naturally but here was assisted by man.

In case the reader thinks that this is simply an isolated phenomenon, it is easy to cite two other chronology-building projects which both observed the same thing. Siebenlist-Kerner (1978, 160) produced two sections of English chronology covering 1341 to 1636 and 1729 to 1969. Obviously it would have been preferable to complete this chronology, but clearly the easily available material did not run across the seventeenth century. Morgan (1977, 813; and personal communication) also constructed two sections of English chronology covering 1359 to 1591 and 1710 to 1974. In point of fact, with the exception of the Nottingham group, using long-lived Sherwood oaks (Laxton *et al.*, 1979, 32) no one in England has definitively bridged the seventeenth century. There seems no doubt that while the dendrochronologists are having difficulty in finding suitable material to bridge this century, there is good reason to believe that we are seeing the results of a depletion/regeneration phase.


## The Fourteenth-century Depletion/Regeneration

The seventeenth-century example is probably not all that serious in the sense that some long-lived oaks do exist in England and Scotland. The phenomenon none the less exists. There is tree-ring evidence to support the historical supposition. The next such event back in time first showed up in the north of Ireland. In Chapter 7 it was observed that the longest-lived timbers from seventeenth-century buildings in Ulster tended to run back to the late fourteenth century. As more and more timbers were dated this picture became clearer and clearer. The oldest timbers used by the seventeenth-century builders had a maximum age of around 250-80 years. It seemed reasonable to suggest that the trees being exploited after 1600 throughout the area of study were the product of regeneration in the late fourteenth century. Now it could be argued that this was merely due to selection of particular sizes of trees by the builders. However, the phenomenon was so general that this seemed unlikely. After all, there must have been some 300- and 400-year-old trees which could have been used, had they existed.

Experience in the Dublin area added slightly to the picture (see Chapter 7). The masses of oak timbers from the Dublin excavations turned out to be mostly thirteenth-century in date. Only one structure pushed forward in time to 1306. Further, in attempting to build a Dublin chronology back to join with the early chronology it was only possible to push back to 1357. Already therefore we had all the ingredients necessary to postulate a depletion/regeneration phase. Chronologies for Belfast and Dublin, taken together with those of Siebenlist-Kerner and Morgan (above), had start years in 1380, 1357, 1341 and 1359 respectively, while Dublin ran out only to 1306. When the maximum extensions of other site chronologies, especially those from Scotland, are added, the picture becomes clear. Figure 11.1 shows the non-continuous nature of tree-ring chronologies across the fourteenth century. One group of timbers which has not been included are the art-historical chronologies constructed by Fletcher (1977). There seems little doubt, however, that these art-historical chronologies are based on rather exceptional trees whose source is at least problematic[1] (see Chapter 12).

The conclusion which can be drawn from Figure 11.1 is that this event or horizon is centred closely on the year 1350. In broad terms there is substantial exploitation of long-lived oaks in the thirteenth and fourteenth centuries and

**Figure 11.1:** The fourteenth-century 'gap' or depletion/regeneration phase appears to centre on the date of the onset of the Black Death. At least three other unpublished English chronologies show this same phenomenon.

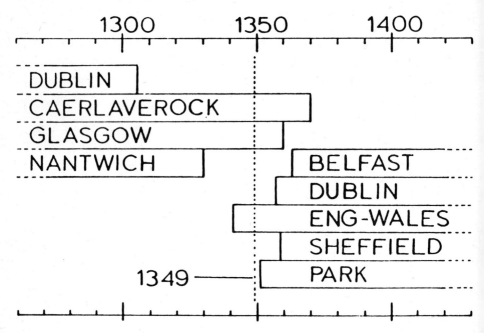

clear evidence for regeneration from the middle of the century. This evidence can be backed up by some circumstantial evidence from the castles at Caerlaverock and Dunsoghley, where in each case timbers likely to have been felled in the early fifteenth century exhibited very short ring patterns, consistent with late-fourteenth-century regeneration.

The coincidence of the central point of this depletion/regeneration with the onset of the Black Death in 1349 is interesting and it is possible to offer a simplistic explanation for the tree-ring observations as follows. The thirteenth century saw a period of expansion both of population and economy. Enormous numbers of trees were felled for a wide variety of purposes, including massive building programmes. There may also have been clearance of marginal land on a considerable scale. In the fourteenth century the overall condition turned to one of decline. Economic problems coupled with exhaustion of the recently cleared marginal land may well have given rise to a stretched population which was particularly susceptible to the ravages of the plagues following on 1349 (Hatcher, 1977). The almost instantaneous reduction of the population by one-third in the mid-century could have allowed the marginal land to be returned to forest at that time.

Clearly the tree-ring gap was not an absolute event in any sense. Long-lived trees were still being exploited as late as 1500 at Lincluden (see Chapter 7). Similarly, it was eventually possible to bridge the gap in the north of Ireland using material from an area where the depletion/regeneration model need not apply (Baillie, 1977b). It is recorded none the less by the maximum extent of the various chronologies in Figure 11.1. We must assume that wherever possible dendrochronologists will attempt to extract the maximum length of ring pattern, i.e. they will try to extend their chronology as far as possible in any instance. What we are seeing in the fourteenth century is a point in time where chronologies consistently stop. Interesting confirmation of this fourteenth-century tree-ring horizon was afforded by the maximum extent of Eckstein's Hamburg chronology which is published back to 1338 (Eckstein *et al.*, 1972). This chronology, with eleven ring patterns beginning between 1338 and 1400, is clearly reflecting a regeneration cycle.

## A Ninth-century Depletion/Regeneration

Having once observed the fourteenth-century gap, it became inevitable that others would be looked for. The first hint of an earlier example was provided by the timbers from the Dublin excavations. As was noted in Chapter 7, the sites in Dublin could be divided into an earlier, essentially non-oak, Viking phase and the later oak-using Norman phase. It became evident that the longest-lived trees used by the Normans had all started growing in the second half of the ninth century. The maximum extent of the chronology was 855. Searching for examples to extend this chronology further back in time proved hopeless. As work progressed elsewhere it became evident that Dublin was not alone in this respect. The Belfast chronology pushed back to 919 and stuck. Scotland yielded a single tree going back to the 890s but overall the chronology was terminated in 946. Timbers from Perth provided a chronology from 949 to 1204 (Baillie, 1981). Fletcher's Ref 6 covered 778 to 1193. Hillam (personal communication) extended a chronology for Exeter to 811.

So over the whole of the British Isles chronologies were extending back to between 778 and 949. The timbers utilised in these chronologies were mostly from long-lived trees exploited in the twelfth to fourteenth centuries. Broadly speaking, from the ninth century oaks were regenerating. It is necessary to look only as far as Germany to see that the original published oak chronologies for west of the Rhine (Hollstein, 1965) and central Germany (Huber and Giertz, 1969) ended in 822 and 832 respectively. It would appear that the phenomenon was not

**Figure 11.2: Although not as dramatic as that in the fourteenth century, there is clear evidence of a nonconformity in the availability of long-lived timbers centred on the ninth century AD.**

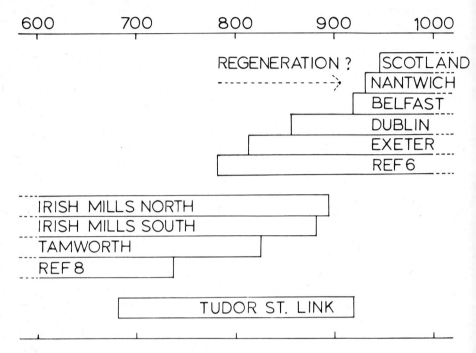

restricted to the British Isles. While this information was building up, effort was being put into the construction of a first-millennium chronology (see Chapter 9). It was hoped that if this floating chronology could be pushed forward far enough in time it could eventually be linked to either the Belfast or Dublin chronology, or both. The final result was the existence of chronologies for the north and south of Ireland running back to 919 and 855 and equivalent 'floating' chronologies running forward to 894 and 881 respectively. This is depicted in Figure 11.2. As detailed in Chapter 9, the tying down of these chronologies was due to the existence of a single ring pattern from Tudor Street, London, which spanned 682 to 918. There can be no doubt that in Ireland in particular there is a ninth-century nonconformity in the availability of long-lived trees. It can be spelt out thus. Up to a date in the vicinity of 930 builders, in particular builders of horizontal mills, had access to long-lived oaks for construction purposes. From that date, for the next two centuries, sampling has not turned up any archaeological timber with anything like 200 rings. It is not until the thirteenth century that long-lived oaks are being consistently used and the longest-lived of these had all started growth in the later ninth century. This event appears to have all the ingredients of a depletion/regeneration phase.

## A Hint in the Early Irish Texts

In Ireland a set of tales referred to as the Ulster Cycle was committed to writing for the first time around the eighth century AD. Previously this had been an oral tradition and could best be described as a quasi-historical mythology. The important point is not the content of the tales but rather the fact that they were written down on a number of subsequent occasions. It is the alteration of subsequent descriptions of the palace of Ailill and Medb which concerns us here. The textual evidence is set down in detail in Mallory and Baillie (1981, forthcoming). Basically the problem is this. In the attested eighth-century description we read of a 'House of oak [with] roof of shingles'. An attested mid-ninth/early-tenth-century rewriting is identical in all major elements with the exception of the 'House of oak' line which now reads 'of pine was built the house'. A still later version of the early eleventh century omits the material of building altogether, using instead 'planks of slats'.

What is interesting is that the dating of these texts had been suggested by Thurneysen in 1912 and Meid in 1967, long before the relevant dendrochronological results became available. In particular the change from oak to pine is striking, as pine is thought not to have been available in Ireland in the first millennium AD. This raises the question of whether pine was being imported, but that is not our concern. Overall what we have here are two completely independent lines of research, textual and dendrochronological, both suggesting a change in the availability (?) of oak in the late ninth/early tenth century AD.

It would be dangerous to suggest that this depletion/regeneration is proven. What can be said is that there exists sufficient evidence to demand further investigation of this period.

## A Post-Roman Regeneration

As soon as the floating chronologies of the first millennium were tied down, it was possible to specify two English chronologies. These were Tamworth, 404 to 825, and Fletcher's Ref 8, 416 to 737. The coincidence of the maximum extent of these two long chronologies may well reflect a regeneration phase following on the decline of Roman influence in Britain. Obviously the evidence is slight as yet, but on the basis of the gaps outlined above, which were each recognised on the basis of a single-area chronology, it is likely that the acquisition of timbers in England to bridge 400 AD may be difficult.

**Earlier Gaps**

Chapter 10 outlined the development of the sub-fossil chronologies at Belfast. The first chronologies constructed were placed in their relative positions in time by radiocarbon dating. New chronologies as they were produced were tested for correlation with the available chronologies and only dated by radiocarbon if no significant agreement could be found. Using this procedure, the original empty time-span was rapidly filled and long sections of chronology were produced. At that stage progress was rapid, since each new chronology had to fit somewhere in time. However, as early as 1975 in the Belfast project it could be seen that 'the law of diminishing returns' was setting in. New chronologies, laborious to construct in time and effort, were found to agree with already available chronologies without extending them significantly. A good example of this was the Charlemont chronology, 900 years in length, which extended the long chronology by only 40 years. By 1975 the proposed 6,000-year chronology was complete with the exception of gaps at 800 AD, 100 BC and 900 BC (in round figures). The example at 800 AD has been considered above. It is likely that the gap centred on 100 BC will yield to archaeological pressure. It is the gap at 900 BC which must be our chief concern. Since the approach used had been successful in completing a chronology for the whole of the first four millennia BC with the exception of the 900 BC gap, the question could be asked: 'Where did the chronology-building process break down?' Initially it was assumed that gaps were simply a sampling problem — collect more and more samples and a bridge is bound to be found eventually. This may be true; however, the situation needs to be looked at in more detail. In the early stages each new group of sub-fossil or archaeological material represented finite progress — as above; each group had to fit somewhere in time. By the time the gap at 900 BC had been recognised, the situation was very different; the search was on for material of specific age. Since all bog oaks and most bogs look alike, this is an extremely difficult task. A slight over-simplification of the problem may serve to drive the point home — somewhere in the bogs, lake-beds or river-beds of the north of Ireland there is a sub-fossil oak which will bridge the gap — but the chances of it being found are extremely slight.

The gap at 900 BC had first shown up in material from Garry Bog, Co. Antrim. This is the largest raised bog in the north of Ireland and over several years large tracts were reclaimed for farmland. Several hundred bog oaks became available and the first groups to be sampled rapidly yielded two chronologies, GB 1 and GB 2, of 1,600 and 700 years respectively. When dated by radiocarbon these chronologies were found to run consecutively, covering the approximate periods 2500 to 900 BC and 900 to 200 BC. Taken at face value, the radiocarbon determinations even suggested that some overlap might exist between the chronologies, although no significant overlap could be found. Since the gap had to be short in either a positive or negative sense, an obvious thought was to collect more material – in theory oaks should have existed continuously on the bog in order to explain their presence after the gap. Extensive sampling was undertaken and the whole length of each of the chronologies was duplicated with fresh material. No trees were found which would bridge the gap, although numerous specimens fell on either side. Various approaches were tried, such as collecting all trees from the bog with more than 300 rings. All these samples either matched the two chronologies or were shown by radiocarbon to be much older. None straddled 900 BC. It began to look as though a depletion period existed on Garry Bog at this date. Various theories – climatic, catastrophic or anthropogenic – were put forward to explain why such a gap might exist, although it was still believed that the most likely cause was sampling deficiency or some artifact of our chronology-building procedures.

For example, as each tree was processed, its ring pattern was compared with the available chronologies using the CROS computer routine. Only statistically significant correlation positions were checked visually, largely because of the volume of material under study, but also because if the match was not statistically significant it would not be acceptable anyway. Was it possible that matching positions at the ends of chronologies were being missed for some reason? No evidence could be found to suggest that the ninth-century BC gap was an artifact of the procedures used. However, it was a source of concern that the chronology building had changed, from cross-matching individuals and constructing site masters, to a situation where individuals were compared with existing chronologies. This latter approach had the disadvantage that individual ring patterns are usually shorter than site chronologies. In order to bridge the ninth century what was really needed was a long site chronology covering the general period 500 to 1500 BC. Since Garry Bog had ceased to hold much hope of bridging this gap, it was decided to search for sites which might yield chronologies of this approximate period.

By and large it is difficult to judge the age of a peat bog without laboratory analysis. However, in the north of Ireland one useful criterion exists which can help to separate sites older or younger than *c.* 2000 bc (radiocarbon years), this is the presence or absence of pine. As noted above, pine dies out and disappears from the pollen record around 2000 bc (Smith and Pilcher, 1973). Thus a bog showing evidence of pine in its upper levels could be ignored, since anything at a lower level would be older than 2000 bc. Two notable successes of this search technique were Ballymacombs More, Co. Londonderry, and Charlemont, Co. Armagh. Groups from both these sources yielded long master chronologies younger than 2000 BC. In each case site masters were produced first, to give maximum chronology length, and these floating chronologies were

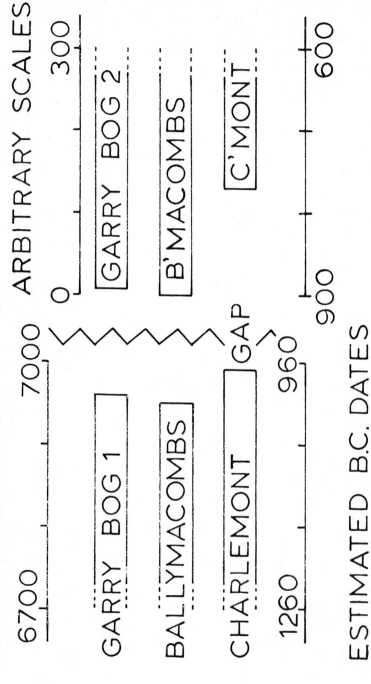

Figure 11.3: The sub-fossil 'gap' presumed on radiocarbon evidence to belong to the ninth century BC. Reference to Figure 10.1(e) shows that a similar gap exists in the German sub-fossil chronologies. If these events are in fact coincident, circumstantial evidence suggests that they are due to increased wetness.

then compared with the existing chronologies. Unfortunately at both sites the GB 1 − GB 2 problem was repeated. Ring patterns ran up to *c.* 900 BC and in each case younger material matched with GB 2 (Figure 11.3).

Since all three sites exhibit a gap at the same time, it would appear fundamentally unlikely that sampling was at fault. There must be a reason for the failure to find a bridge − there must be a real gap or depletion period at that time. Now obviously material of this age must exist; oaks did not die out at 900 BC, but the problem of finding it is more complex. Not only must the search be restricted to sites of a particular period, but there is the added stricture that the source must be one not affected by whatever caused the gap at Garry Bog, Ballymacombs and Charlemont. Since the cause is not known on these three sites, this problem is extremely difficult.

One possible solution to this gap would be to explore sources of material outside the north of Ireland. This, however, brings in the added complication that the ring patterns might not match with those from the north of Ireland. It is known from modern and medieval studies that cross-matching can be obtained between chronologies from Ireland and England. However, it is also known that there is little or no matching directly between Ireland and Germany where the nearest equivalent oak chronologies exist. There appear to be two lines of approach to the consolidation of these Irish chronologies. One might be by a stepwise link with Germany via England. That is, a replicated floating chronology could be constructed for some area in England. Suitable timbers certainly exist in western England in the fenlands of East Anglia and at various points around the coast of England and Wales (Heyworth, 1978, 279). Hopefully the English chronology could then be matched against both the Irish and German chronologies. The success of such an exercise could not be guaranteed, but an attempt would certainly seem worth while.

The alternative is to wait for the chronologies to be tied down by the high-precision radiocarbon dates performed for calibration purposes. This should allow the real age of the floating chronologies to be specified to within ± 10 years (Pearson, personal communication) (see Chapter 12). One significant point in this hypothetical teleconnection between Ireland and Germany relates to the actual date of the so-called 900 BC gap. Pearson suggests that, on the basis of correlation between the detail of the Belfast calibration and that of Suess, the 2,990-year Belfast long chronology ends around 890 BC. The Garry Bog 2 chronology of 700 years presumably lies between 100 and 890 BC. So the 900 BC gap is most likely to occur in the ninth century BC. If we refer to the work of Becker and Delorme (1978, 61) we discover that the German precisely dated chronologies have a maximum extent back to 700 BC. The next floating chronology back in time is dated to 800 to 1300 BC on the basis of calibrated conventional radiocarbon dates. It is not impossible within the limits of accuracy of the present discussion that the German and Irish chronologies are both exhibiting a gap at the same point in time. Resolution of the dating problems should show within the next few years whether or not this is the case. If both projects *do* show a synchronous gap the causal mechanism could be an interesting one. (See also the section on geomorphological dating in Chapter 12.)

### Gaps: the Corollary

One point should be clarified. No one set out to look for gaps. Life would have been much simpler had a continuum of readily available timbers existed. The gaps are there, however, and show most strongly in the fact that they appear in the results of separate workers. So we can suggest that there are periods from which relatively large numbers of long-lived timbers survive separated by intervals when such timbers virtually do not exist.

The corollary of this is that when a long-lived timber of the last two millennia is encountered, it is not free to be of any date. It is fairly heavily constrained to fall between the gaps. In Britain such a timber must tend to fall between 100 BC and 400 AD, 400 and 850 AD, 850 and 1350 AD or 1350 and 1700 AD (hopefully a 300-year modern tree would be recognised as such). In Ireland, where lack of Roman involvement nullifies the 400 AD gap, the first two may be lumped together.

### Note

1. Fletcher suggests that the oak boards came from the stems of high forest trees which were specially selected and states 'The pattern of MC18 is therefore derived from a small and unrepresentative fraction (perhaps no more than 0.1%) of the oak timber grown at that time in the region' (Fletcher, 1977, 337). How the manufacturers of oak boards could have selected such an unrepresentative group is not clear since they cannot have made the selection on the basis of ring patterns.

*Chapter* 12

# Applications

The preceding chapters have outlined in some detail the development of basic reference chronologies. Such exercises are necessary to establish the technique. However, dendrochronology as a straightforward tool for the dating of buildings and archaeological timbers can be supplemented by a variety of other applications. In this chapter it is intended to look briefly at some of these. Probably the most important from the point of view of the archaeologist is the close tie between dendrochronology and the calibration of the radiocarbon time-scale. Also important is the dating of less well localised timbers than those used to develop chronologies. In particular this relates to the timbers of ships and boats whose origin may not be known. Third, to redress slightly the balance towards pre-historic dating there are some things to be said on the subject of relative dating.

There are of course other areas of study, such as climatology, geomorphology and art history. which can be aided by dendrochronology. These are largely outside the scope of this book but nevertheless require mention.

## Radiocarbon Calibration

For archaeological purposes radiocarbon is undoubtedly the most important scientific dating method. Its appeal lies in its universal applicability. Essentially any organic remains of the last 70,000 years, and potentially of the last 100,000 years, are open to investigation. The method in outline is as follows. Natural radioactive $^{14}$C is produced only by the bombardment of the upper atmosphere by cosmic radiation. Initial bombardment produces slow neutrons which in turn interact with nitrogen nuclei to produce $^{14}$C. The $^{14}$C enters into a 'fixed' proportion with stable $^{12}$C in atmospheric carbon dioxide. Large amounts of dissolved carbonates in the world's oceans act as a buffer to extreme variations in the $^{14}$C content. Since living tissue fixes its carbon from the atmosphere, the result is a fairly constant proportion of $^{14}$C: $^{12}$C in plants and animals. After the death of an organism the balance is no longer upheld and the proportion of $^{14}$C starts to diminish. Since the rate of decay of $^{14}$C is known, it is possible by

223

measuring the proportion of $^{14}$C: $^{12}$C in an ancient sample to estimate the time elapsed since death.

The method works extremely well if all one wants is a broad estimate of age. Problems arise as soon as attempts are made to establish dates. There are two main reasons for this. The first relates to the inherent error associated with the laboratory measurement of the amount of $^{14}$C in a sample (the activity). This inherent error tends to smudge the age determinations so that no routine radiocarbon date can be guaranteed within a century either way. Individual measurements can in fact be much worse than this. However the second problem is more fundamental. It relates to the basic assumption, implicit in the method, that the cosmic ray flux has always been a constant, i.e. that the $^{14}$C has always been produced at a steady rate.

It was initially shown by De Vries that this assumption was incorrect and that the amount of $^{14}$C in the atmosphere did vary with time. Archaeologically the non-constancy of $^{14}$C was most clearly shown in the measured activities of 'known-age' Egyptian samples. The activities for such samples before 500 BC were too high compared with their expected activities, giving radiocarbon dates too young by the order of hundreds of years. In order to check this discrepancy definitively it was necessary to date large numbers of known age samples. The bristlecone-pine chronologies offered just such an opportunity. By measuring the activities associated with successively older blocks of precisely dated tree-rings it was possible to provide an empirically derived relationship between $^{14}$C activity and real age, a calibration curve.

This opportunity was exploited in the late 1960s principally by Suess, who in 1970 produced the first calibration curve for the interpretation of radiocarbon dates. This calibration showed that while Libby's theory was broadly correct, it was not precisely so. The calibration did explain the discrepancy between the Egyptian historical dates and the raw radiocarbon dates. However, the story did not end there because several questions remained to be answered. For example, was this American calibration applicable throughout the Old World (McKerrell, 1975)? Remember that the bristlecones grew at high altitudes in the western United States and might be atypical in some way. Second, was the fine detail of the Suess calibration (i.e. the wiggles) real or apparent? Broadly the world divided between the Suess school which advocated wiggles and the statistical school which plumped for smoothing of the calibration. This smoothing was a reaction, on the part of statisticians, to the inherent errors in the individual measurements which were felt to be underestimated.

It was as a reaction to just these questions that the work at Belfast on the construction of a long independent tree-ring chronology was begun. Using Irish oaks, the possibility existed of checking Suess's results with a low-altitude, Old World calibration. However, while work was progressing on the construction of a long oak chronology some fundamental reappraisal of the question of $^{14}$C measurement was in progress. Basically, the problem was this. Routine radiocarbon dates involved measurement of sample activity. The error associated with the resultant date was based on the counting statistics of the activity measurement. It had been assumed that to decrease the error all that was necessary was increased counting time. However, workers such as Stuiver (1978, 271) and Pearson (1979) realised that the advantage gained through increased counting time was at least partly lost in long-term counter-instability. They took the line that to reduce realistically the errors associated with radiocarbon activity measurement, and hence with radiocarbon dates, specialised highly stable set-ups would be necessary. In addition, lower backgrounds and generally tighter procedures would have to be instituted throughout the measurement process. Thus the later 1970s saw the development of a small number of new high-precision set-ups. In comparison with the normal routine dating facilities around the world, these high-precision units represent what might be called second-generation radiocarbon machines. The reality of the situation is as follows. If routine machines could produce dates with standard deviations no better than ± 70 (even where lower figures were quoted it is likely that ± 70 would be a more realistic figure), the second-generation machines were aimed at genuine precisions of ± 15 or better.

## The Second-generation Calibration

One of the snags encountered by Suess related to the very narrow rings of the bristlecones. Ferguson, who processed the samples, could only supply a small number of grams for each ten-year block of rings. A major advantage of European oaks was their relatively wide rings and hence the relatively large samples available. At Belfast, Pearson designed his high-precision set-up around a sample size of 175 g for each block of 20 rings. Similarly, Becker has been able to supply other laboratories with large samples derived from German sub-fossil oaks.

The final results of the high-precision calibration will not be fully available for a number of years. The intention is that in the final definitive version, each section of chronology should have been measured independently by at least two high-precision laboratories. To demonstrate the degree of similarity between results on different samples measured at different laboratories, Figure 12.1 shows the last 500 years as analysed by Stuiver and Pearson. Clearly there is fine detail in the curve and hence real short-term variation in the $^{14}$C concentration in the atmosphere with time (Stuiver, 1978; Pearson, 1981).

In broad terms, between the work of Stuiver, Pearson and De Jong almost the whole of the last 7,000 years has been calibrated at high precision. One thing has become apparent from this work. The results produced by Suess do show the same general details in terms of periods of enrichment and depletion, i.e. the Suess wiggles were broadly correct. However, his results suffered from the noise

**Figure 12.1:** The relationship between conventional radiocarbon ages (high-precision) and tree-ring age for the last 500 years. Open circles represent measurements by Stuiver, closed circles by Pearson.

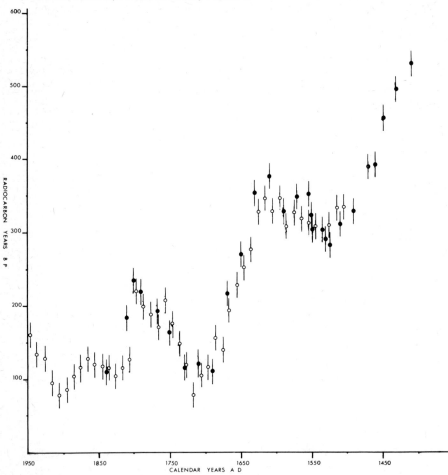

Source: Courtesy G.W. Pearson.

associated with routine dating and as a result there is a greater spread on his data than on that now becoming available (Pilcher and Baillie, 1978; De Jong *et al.*, 1979).

This last comment brings us back to an important point with regard to routine radiocarbon dates. While the high-precision calibration, when it becomes available, will finally resolve our understanding of the calibration relationship, it is not going to improve the interpretation of routine dates radically. The reason is that each routine date will still retain its broad spread of probability. Figure 12.2 shows how that spread comes through the calibration procedure to leave the archaeologist with at least as great a spread (in some instances a much greater spread) of real time.

**Figure 12.2: Radiocarbon dates tend to become smeared in the calibration process. Even using a high-precision calibration (after Pearson, 1980) a routine radiocarbon date with $\sigma = \pm 70$ can easily expand its range from 280 to over 400 years at 95% confidence limits.**

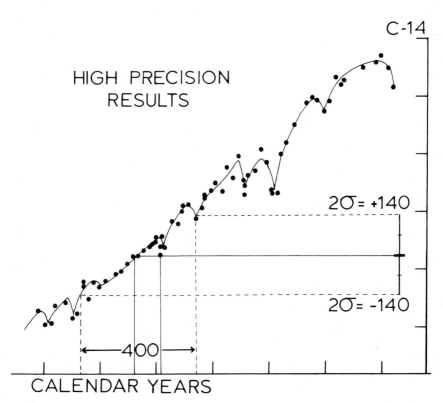

One obvious suggestion has to be that the high-precision calibration will be most suitable for calibrating other high-precision dates. This is true, but there is no room for euphoria. The very fact that there is significant detail associated with the high-precision calibration, as in Figure 12.1, means that there may well be ambiguity, even with individual high-precision measurements. Consider a high-precision $^{14}$C date of 350±15 bp. This could derive from a sample which lived either around 1480 or between 1550 and 1650. Clearly there could be no definitive interpretation of this date. This implies that to use the potential of high precision to advantage archaeologists are going to have to supply samples which are capable of subdivision — for example, pieces of wood or charcoal with sufficient rings to allow the equivalent of three or more consecutive calibration samples. Only in this way will definitive interpretation of the results be possible. In fact, the method at this level of sophistication is going to require a mini-calibration for each sample, but will pay off with dates which may be accurate to within ten years in real time.

The corollary of such a situation is that archaeologists are going to be required to supply samples with much greater integrity to justify such accuracy. With the likely cost of a set of high-precision dates being in the hundreds of pounds only really relevant samples may be considered. All those awaiting the high-precision calibration as the final answer to all questions of interpretation be warned!

### Specifying the Age of Floating Chronologies

One immediate advantage of the presence of fine detail on the definitive calibration and on Suess's curve is that the detail can be compared. This opens up the possibility of tying down the sections of floating European calibration against the fixed bristlecone-pine calibration. It has to be remembered that both the Irish prehistoric chronologies and the German chronologies older then 700 BC are floating (see Chapter 10). Since each is being calibrated at high precision it should be possible to cross-match the detail on the calibrations both with each other and with those of Suess. This should allow specifications of the real age of the floating chronologies to within ± 10 years (Pearson, personal communication). Once this has been accomplished the dendrochronologists can compare the actual tree-ring chronologies to see if there is any agreement. However, it is recognised that such a procedure can be dangerous, because after all if a match exists it should be possible to find it independently. It is likely, however, that no direct cross-correlation exists between Ireland and Germany. The answer may be to construct some, even relatively short, sub-fossil chronologies in England. Hope-fully, as noted in all of the later periods (Chapters 6, 7 and 9) such stepwise agreements might allow linking of the Irish and German chronologies. Such efforts may allow the eventual completion of both chronologies and of the European high-precision calibration.

### Conventional $^{14}$C Dates for Known Age Samples from the British Isles

One side-effect of the tying down of the Belfast early medieval chronology and the subsequent dating of a number of English chronologies has been the speci-fication of a number of radiocarbon samples in real time. Although only the most general account had been taken of radiocarbon dates for the placement of wood samples in time, quite a number of such dates had been obtained, mostly on samples about which very little was known. With the chronologies tied down, the radiocarbon dates can be plotted against real time as a calibration. Figure 12.3 shows the spread of results along with those recently published in *Radiocarbon* by Bruns *et al.* (1980, 276). Clearly the routine dates are much more widely scattered than the more recent high-precision dates, the latter having standard deviations around ± 20 years compared with ± 70 for the former. From the point of view of the archaeologist who is regularly dealing with just such routine dates, these results are sobering. If we accept that the Bruns results represent the real relationship between radiocarbon and calendar dates, then the standard deviations on the routine measurements are genuinely of the magnitude suggested above. In a word, routine measurements vary over a range of about 300 years (this would be consistent with the 95 per cent confidence limits on dates with standard deviations of ± 70 years) and attempts to interpret individual dates or even small groups of dates without taking this into account may well lead to totally false conclusions. The sad fact, and one to which archaeologists must become accustomed, is that calendar dates cannot be reliably deduced from *routine* radiocarbon measurements.

## Geomorphological Dating

It is obviously of considerable interest to geologists and others to know the rates at which natural events occur and the dates of specific incidents. Dendrochronology can on occasion supply such information. In some instances the presence of trees of a particular age in a deposit is sufficient elucidation. In other cases some physical change within the tree may give the clue to the dating of an event. In Chapter 11, in the construction of the Irish sub-fossil chronologies, each site chronology tells us the length of time a particular bog was capable of supporting oak growth. In addition, from the age spectrum of a site chronology it is possible to infer whether the timbers were part of a regenerating forest or a once-only colonisation. The distribution of the end years of the various trees tell something of the nature of their deposition. Now all that information falls naturally out of the dating of a random sample of oaks from a bog. With stratified remains, despite the cautionary remarks of Chapter 10, it is sometimes possible to trace a more detailed history.

In Germany the extensive sampling of sub-fossil oaks from river gravels has allowed the detailed study of the development of the valleys of rivers such as the Main, Rhine and Danube. By dating oaks in ancient silted river courses both the movements of the river and the times of increased fluvial activity can be studied.

**Figure 12.3: Comparison of the high-precision dates presented by Bruns *et al.* in *Radiocarbon*, vol. 22 (1980, 276), where each dot has σ = ± 20, with 20 routine dates (as used by archaeologists) on known age oak samples.**

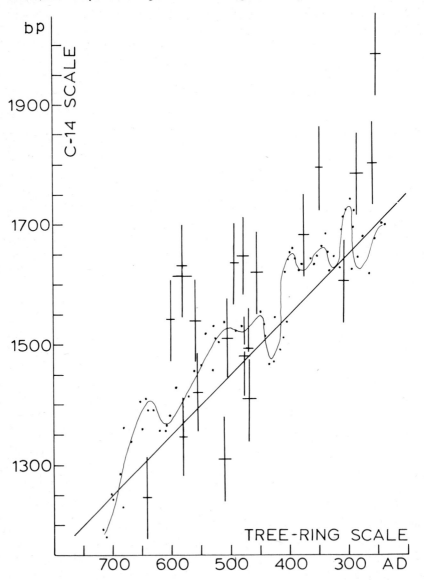

Source: Routine dates courtesy G.W. Pearson with additional Harwell dates reconstructed after Fletcher (1977).

While in Irish bog contexts the oaks are normally recovered from the same site on which they grew, with river gravel oaks the original context is lost and the interest lies in the period of deposition. In river gravels the distribution of the end years in a group is of particular interest. Regular deposition over a long period infers stable conditions, while distinct phases of increased river activity can be reconstructed from any significant clustering of cross-dated trunks.

On the Danube the main phase of oak deposition took place between 2210 and 1290 bc (radiocarbon years), with around 130 trees represented. Interestingly, the second-largest group of oaks from the River Main was desposited between 1800 and 1450 bc (40 trees). This shows that an erosion phase was occurring in both river systems simultaneously. Following this active phase in the Bronze Age there was a phase of apparent inactivity until a few centuries BC. During that period only one trunk layer is known from the River Main. This group, M15, washed out between 730 and 640 bc (radiocarbon years). It comprised trunks with an average age around 300 years and this implies a purely local interruption of an otherwise undisturbed riverine forest development (Becker and Schirmer, 1977, 307).

This last observation is of some interest with respect to one of the 'gaps' in the Irish long chronology. In Chapter 11 it was shown how trees on three bog sites in the north of Ireland occurred on both sides of a gap which could be dated on the basis of high-precision calibration measurements to around 890 BC (real years). One possible cause of this nonconformity was increased wetness. This seemed the most likely cause given the low-lying nature of the sites. Now we see an independent event in a German river valley whose *uncalibrated* radiocarbon age is 730-640 bc. It is just possible that we are in fact seeing two aspects of the same event, namely a short period of increased river activity (washing out trees which had been growing uninterrupted for centuries) in Germany at a time when increased wetness (?) was interrupting the growth of oaks on Irish bogs. Certainly a point worth looking into. While on the subject of gaps, it is worth noting that Becker's largest River Main deposition period begins in 226 BC (real years). Is it possible that this period of increased river activity is associated with the final disappearance of oaks from Garry Bog, as indicated by the end of the 700-year GB 2 chronology (see Chapter 11).

## Internal Evidence from Tree-rings

Similar exercises can and have been carried out in any context where trees are preserved. Of equal importance, however, is evidence for such events as landslides, earthquakes and volcanic activity preserved in anomalous ring patterns. At the simplest level, if a tree is damaged by fire, but not killed, it may preserve evidence of the year of the damage as scar tissue within its trunk. Fritts (1976, 221) illustrates a pine section with evidence of fire damage on four separate occasions in 1839, 1859, 1868 and 1918. On the other hand, a tree whose orientation is changed by eroding ground surface, changing valley profile or more violent earth movements may well exhibit a dramatic change in the symmetry of its stem to the formation of tension (in angiosperms) or compression (in gymnosperms) wood.

One example of the dating of volcanic activity is furnished by the trees near Sunset Crater, Arizona. This volcano is situated in a semi-arid area of north-central Arizona and vegetation is only gradually recolonising the immediate locality. Sunset Crater represents the most recent event in the San Francisco Mountains volcanic field, which contains evidence of about 400 extinct volcanoes. The debris of the eruption consists of lava flows and extensive ash layers. Living trees, several hundred years old, are rooted on top of these deposits. How could the date of the eruption be specified? In the first instance it could be narrowed down on the basis of archaeological evidence. Archaeological sites existed both under the ash layers and subsequent to their deposition. As usual in semi-arid sites, timbers were preserved and these could be dated against the chronologies constructed by Douglass. The latest pre-eruption date for any tree-ring specimen was AD 1046, while the earliest post-eruption date was 1071. So conventional dendrochronology could narrow the date of the eruption to a quarter-century. The actual date of the eruption was eventually specified by a group of timbers from a ruin at Wupatki. This site lies about 19 km north-east of Sunset Crater and contained timbers felled in the early twelfth century. These trees showed dramatic curtailment of growth following on the ring for 1064 (Smiley, 1958, 190). Thus the eruption must have taken place late in 1064 or early in 1065 (see Figure 12.4). In order to give some perspective on an event which may have physically damaged trees up to 19 km distant, it is estimated that the incident threw up something in the order of 1,000 million tons of lava and ash.

## Relative Dating

We have been considering dendrochronology almost exclusively as an absolute dating method. Certainly that is where the method is at its most elegant. There are, however, circumstances where dating exercises which fall short of absolute are still acceptable. There are three main areas where relative or internal dating is infinitely preferable to no dating at all. Broadly these apply to groups of timbers of an isolated species, to sites which are physically isolated and sites which are isolated in time.

It has already been suggested that absolute dating for the prehistoric period in the British Isles is something of a non-starter (see Chapter 10). The shortage of timber-bearing sites older than the first millennium BC would make the building

**Figure 12.4: Severe set-back in pines from the Wupatki ruin specifies AD 1064 as the year of eruption of Sunset Crater.**

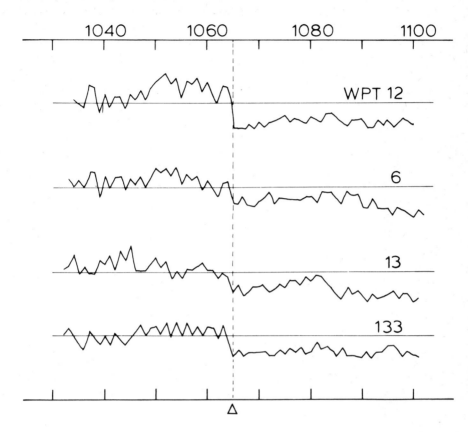

Source: Smiley (1958).

of long sub-fossil chronologies in Britain futile if their only aim was archaeological dating. Remember that the Belfast long chronology effort was aimed at radio-carbon calibration. At Belfast we have an almost complete 7,500-year chronology and no serious timber-bearing sites older than the Iron Age. The logical outcome of this line of thinking is to divert attention away from absolute dating – a concept which like the high-precision calibration seems to be a Holy Grail for some archaeologists – and place a good deal more emphasis on internal site analysis should the opportunity arise to study occasional timber-rich sites. This is particularly true of prehistoric sites, but may ultimately take on greater importance for sites of more recent times. Basically, there is an obsession with absolute dating as a concept, but the real irony is that archaeologists are seldom geared to cope with such results when they do become available.

The point is that the dendrochronologist is in a unique position *vis-à-vis* the archaeologist. Let us consider a hypothetical example. A site is under excavation, and most consideration and discussion will be devoted to associations or typology and the question of what date the site is. Most archaeological reports have lengthy discussions on the likely date of a site or structure and the reasonings behind the choice. What happens when the dendrochronologist is able to say, for example, that the Drumard horizontal mill (Chapter 9) was a one-off structure built using timbers felled in AD 782? What happens is that all the previous discussive evidence on date compresses into a single sentence. Dendrochronology is in fact a sure conversation stopper.

So what we may be coming round to is that the absolute date is not all that important in many cases. Certainly it is important at sites like Caerlaverock Castle (Chapter 8) which have documentary associations, but more often there is other information which may be of greater actual importance. Let us consider the Drumard example in this light. We know its absolute date, but there is little we can do with it (with the exception of comparing it with the dates of other similar sites). Our interest should really lie in questions such as 'Were horizontal mills built in a single year? How long was such a mill in operation? If there were rebuilding phases, at what intervals did they take place? . . . etc.' At Drumard it was possible to show that the main structural timbers were all of the same date. Moreover, the sole tree — a horizontal timber which supported the horizontal mill-wheel and formed one of the most important elements in the mill mechanism (see Appendix 2) — could be shown to be of the same date as the structure. So the lifetime of the mill was represented by the length of time necessary for the revolving wheel to wear three holes through the 5 cm thick oak plank (Baillie, 1975). This information would appear to be of considerably more importance than the date.

The important point is that this type of within-site analysis is possible whether or not a chronology is available to precisely date the whole assemblage. It might be justifiable to suggest that with prehistoric sites in the British Isles relative information has to be of greater importance than absolute. As a second example, take the ash timbers from the Dublin medieval excavations (Chapter 7). It would be impossible to build an independent ash chronology back to the eleventh century to join with, and hopefully date, the ash on the site. However, current work at Belfast is showing that good cross-dating can be obtained between ash timbers from eleventh-century pits, houses and corduroy pathways. So a relative scheme is entirely feasible. In the first instance this can be tied down on the basis of archaeological evidence, probably to within a quarter-century. Conceivably it may be possible to cross-date the ash chronology with the existing 855 to 1306 Dublin oak chronology, but of course there is no guarantee that such dating will be possible. However, in this case the absolute dates are not so important as the potential information on rebuilding intervals, structure lifetimes and site development generally. All of this can be provided by relative dating without worrying about the absolute dates. Similarly, reference to the 1960s work at Novgorod in Chapter 1 shows that the relative information on building phases was by far the most important contribution to the understanding of that site. The following are some examples of relative dating exercises.

## Pazyryk Burials

These wooden tombs or Kurgans were located in a small steppe valley in the eastern Altai and were excavated in 1929 and the late 1940s. Chronologically they belong to the fifth to third centuries BC. A number of timber samples were obtained from each Kurgan in the group of five. All of the timbers were of Siberian larch (*Larix sibirica*). The information came at two levels: the first internal to each Kurgan, the second the relationship between the group. For example, Kurgan 1 yielded ten samples, from the walls and roof of both the inner and outer chambers and from the 'dead floor'. All had been felled in the same year and the most likely season was the early spring. There were similar findings from the other Kurgans, sometimes with re-used timbers showing up. The overall dating spectrum came out as follows:

Kurgans 1 and 2 dated from the same year;
Kurgan 4 was constructed seven years after 1 and 2;
Kurgan 3 was constructed 37 years after 1 and 2;
Kurgan 5 was constructed 48 years after 1 and 2

So this remote group of burials yielded some useful information on the apparent life spans of the chieftain class in the area at a few centuries BC. Absolute dates would only have any meaning in this context if the possibility existed of identifying who the individuals were (Zamotorin, 1959).

## The Rhone Bridge at Cologne

This is an example of the way in which relative dating can catch the investigator out. It applies when a series of structures is dated relatively and the whole assemblage is dated on the basis of one site only. Obviously if the dendrochronologist is correct in his dating there will be no problem. However, if he has the misfortune to 'get it wrong', quite often some of his relative dates will turn out to be impossible. Remember that this was just what was suspected when the 'Shaw's Bridge' timbers threw doubt on the Belfast chronology in Chapter 6.

When Hollstein was constructing his 2,700-year chronology for western

Germany, he arrived at a situation where he had an 855-year floating chronology spanning the fifth century BC to the fourth century AD. This was constructed principally with oaks from Roman and Celtic bridges and material from the salt mines at Bad Nauheim. Now, on the basis that the last ring of eight timbers from the Roman bridge at Cologne belonged to the year 310, Hollstein (1978, 41) specified the chronology to the years 486 BC to 369 AD. Now the snag was that once specified, this also fixed a series of earlier Roman timbers which cross-matched with the 855-year chronology. The result was to give the Roman scholars something to get their teeth into. First, it became clear that the Romanists had a lot less faith in the documentary date of the bridge in question. Second, they pointed out that Hollstein's eight timbers came from a population of something like 1,500 posts in the bridge as a whole. This lessened the chances of his having sampled original timbers. The crunch came when Hollstein dated timbers from a large rectangular structure associated with the south-east corner of the Cologne wall. This he dated to 24 BC against his specified chronology. At this point the Romanists pointed out that in 24 BC there was only an oppidum in the vicinity, not a full town with which such a structure might be associated. Basically what they were saying was: 'You have dated a Roman structure to 24 BC, a full decade before the first legions even entered the area' (Baatz, 1977, 178). Baatz in fact suggested that Hollstein's chronology was too old by something like 30 to 70 years. As it turned out, the dating of the Cologne bridge was in error by 26 years, the chronology, as predicted by Baatz, having to move forward from 369 to 395 AD (Hollstein, 1979, 74). The lesson here is: do not specify your floating chronologies until the dating is definitive.

### Cullyhanna Hunting Lodge and Imeroo

There is a separate area of study where relative dates can supply food for thought. In Chapter 10 the dating of the enigmatic hunting lodge at Cullyhanna, Co. Armagh, was mentioned. This site is the nearest thing we have in the British Isles to a precisely dated prehistoric site, close to 1500 BC. Remember that the finds totalled four flints. All that is known about the site is that it consisted of a hut and a palisade, presumably beside (as opposed to in) Cullyhanna Lough (Hillam, 1976).

In 1978 a stray oak timber, QUB 3471, turned up on the edge of a bog at Imeroo, Co. Fermanagh. Since it was a worked timber it was assumed that it had come from a crannog and attempts were made to date it against the chronologies of the last two millennia. No matching position could be found so a sample was submitted for routine radiocarbon analysis — in the hope that the timber might fall in one of the 'gaps' at 100 BC or 800 AD. When the date became available, *c.* 1300 bc (radiocarbon years), it was realised that we had been looking for a match in the wrong place. The 236-year ring pattern of QUB 3471 was then compared with the Belfast long chronology. A matching portion was found with the outer ring equivalent to 6391 on the computer scale ($t = 6.1$).

Now 6391 on the computer scale is equivalent to *c.* 1500 BC. In fact the Cullyhanna chronology ends at 6375 on the computer scale. Subsequent investigation at the Imeroo site suggested that we were dealing with a habitation site of some kind (as yet unexcavated) situated beside what is now a bog but which in the past may well have been open water. Exposed sections in the bog adjacent to the site indicated clay layers which must have been the result of flooding episodes in the past. So now we have two individually fairly enigmatic sites. However, the coincidence in their dates − allowing for missing sapwood on the Imeroo sample suggests felling in 6423 ± 9 (remember this is a computer scale which runs in the opposite direction to the real dates), while the Cullyhanna timbers were felled in 6375 − may well be suggestive of some phase of activity. As noted elsewhere, phases in dendrochronological studies tend to show up at an early stage, for example the clustering of crannog dates noted in Chapter 9. On this basis it does seem a little strange that in the whole spectrum of Irish prehistory the first two sites to be tied down by tree-rings fall within 50 years.

## Nautical Applications of Dendrochronology

Obviously it should be possible to date the ring patterns of timbers from ships and boats by exactly the same procedures as have been applied to timbers from buildings or archaeological sites. The one proviso is that a relevant area chronology should be used. Here of course is the snag. In many instances, especially when dealing with ship's timbers, the area of origin may not be known and need not be the same as the area where the remains are found. If the area of origin is known, the question becomes trivial and ships' timbers may even be used in the construction of area reference chronologies. An example of such usage can be found in Barefoot's Winchester chronology where timber from HMS *Victory* provided an extension to a living-tree chronology (Barefoot, 1975, 25). With nautical archaeology in general the problem is in assigning an origin to any given remains and the following statement from Farrell and Baillie (1976, 45) adequately sets the scene.

> In nautical archaeology one of the major problems confronting the excavator of ancient ships' timbers is identifying where any one excavated ship was built. This point has been stressed in an important and timely article written by Lucien Basch. Referring to the 'inherent limitations' of submarine archaeology, Basch writes, 'The most important limitation, in my opinion, is the virtual impossibility of deducing the shipyard where a vessel was built'.

George Bass encountered the same problem after excavating a Bronze Age wreck off Cape Gelidonya, Turkey. In a report of the excavations of the Kyrenia wreck, one of the excavators, H.L. Swiny, wrote, 'The identification of where an ancient ship was built is very difficult. One might hope for a name-plate or perhaps a rare wood used in the construction which might delimit the area'. It is a difficulty that will recur as long as wrecks and ships' timbers continue to be discovered.

Now from a tree-ring viewpoint this allows an interesting line of reasoning. If on an archaeological site timbers occur which can clearly be demonstrated to have originated from a ship or boat, they can either be local or exotic. If they are exotic, the question which is as important as their placement in time is their area of origin. If a situation arises where the approximate period of the vessel can be deduced from archaeological evidence, it is then possible to compare the ring patterns from the vessel with all available chronologies for the period. In theory, the chronology against which the best definitive cross-dating can be demonstrated should represent broadly the area of origin. This last statement needs to be qualified to some extent. Not enough is known about regional variations for a categoric attribution to area, but the results may well be highly suggestive.

Just such a situation arose in the study of the timbers from the Dublin excavations (Chapter 7). In the construction of the original Dublin 452-year chronology, all timbers which showed signs of re-use from ships or boats were ignored because of their possible exotic origin. However, by 1977 chronologies covering the tenth to thirteenth centuries were available for Scotland, Belfast, Dublin, south-east England and Germany, and each ship-derived ring pattern could be compared with all of these. Obviously it would have been nice to show that some of the timbers were imports. Unfortunately the results were not as one would have expected. The majority of ship or boat timbers tested showed clear agreement with the Dublin chronology (Baillie, 1978b, 262). (Those which did not match definitely against the Dublin chronology did not match any of the other chronologies.) Of particular interest was the much higher level of agreement with Dublin compared with the levels against the other chronologies from the Irish Sea basin. While it would have been more elegant to demonstrate an exotic origin, the results obtained none the less demonstrated that a degree of localisation was possible. Of twelve timbers which matched best with the Dublin chronology, all, with one exception, produced $t$ values in the range of 4.6 to 10.2.

One possible solution to the local nature of the Dublin ships' timbers may of course have lain in the types of craft represented. Since the timbers occur basically as disjointed planks, it could be difficult to reconstruct what sort of craft was being dealt with (McGrail, 1978, 241). Wallace's suggestion that the river Liffey was 'much more broad and shallow in the twelfth century than it is now' implies that indeed ships had to anchor at a distance from the port and probably used lighters (Wallace, 1976, 31). He also reports the particularly relevant petition to Edward III from the Dublin merchants (1358) 'from want of deep water in the harbour . . . there never had been anchorage for large ships from abroad'. This may explain the local finding above. The vessels being dated were only small craft.

Subsequently it was possible to make a further test on timbers from what appeared to be a *ship* from the Dublin excavations. These timbers were figured by McGrail (1978, 242) and were estimated to have come from a ship some 20 to 25 metres in length (Farrell, personal communication). Disappointingly, all the planks studied from this ship also dated directly with the Dublin chronology, this time with $t$ values from 9.6 to 13.7. In this particular case one of the timbers, QUB 3137a, still retained some traces of sapwood and the construction date could be estimated at 1195 ± 9. This implied a ship of Norman rather than Viking construction (the Normans having taken over in the 1170s). Putting the date and the likely local origin together suggests the construction of substantial vessels at Dublin in the later twelfth century.

It is unlikely that all ships' timbers found in excavations will be local in origin. Therefore even these initial results from Dublin suggest that the basic idea is workable. As the number of available area chronologies increases, it may be possible to achieve quite a high degree of localisation. One final cautionary word is necessary. Although the timbers discussed above have been attributed to Dublin, this does not preclude the possibility that, in the future, as chronologies from the west of England become available, they may prove to be even more similar to some adjacent area. It can be stated with certainty that they are unlikely to have derived from continental Europe.

## Dug-out Boats

Dug-outs or log-boats are likely to have been in use within the British Isles from the Neolithic period at least. Although their area of origin is seldom in doubt, there are some problems with the dendrochronological dating of these vessels. By their very nature they tend to be hollow and for most of their length they are represented by a mere shell containing relatively few rings. At the ends, provided that they survive, it may be possible to measure most of a radius. Overall, most log-boats are in less than ideal condition.

Let us assume that in a given case it has been possible to measure the ring pattern of a log-boat. One question which immediately arises is the general period to which it belongs. Inevitably it has to be compared with all the available chronologies for the area. Here is a further snag: usually there will be only one ring pattern available. So it is not possible to create a master chronology for the

**Plate 10: Partly finished dug-out from Oxford Island, Lough Neagh, belonging to the early sixth century.**

Source: Photo courtesy Ulster Museum.

vessel as one could for a clinker-built boat. An example of the dating of a clinker-built boat is provided by the Blackwater Boat, found by Steve Briggs in dredgings from the river Blackwater, where it flows into Lough Neagh. The remains in this case consisted of the keel and six oak planks. The timbers allowed the construction of a 215-year site master which cross-dated with the Belfast chronology with its outer ring equivalent to 1661 ($t$ = 9.1). Allowing for missing sapwood, the boat was constructed in the range 1693±9 (Baillie, 1973b, 26). With the log-boat only a single ring pattern is available and as a general rule the dating between a single ring pattern and a master chronology will not be as good as that between two masters. So the ring pattern has to be tested over a long spectrum of time and its chances of dating definitively are not all that high. This means that in the case of anything other than the most definitive matching, a radiocarbon determination will be necessary to reinforce the date arrived at. Since a radiocarbon date is more than adequate for most dug-outs, in most cases that should be the first approach.

However, consider a case where direct dendrochronological dating was possible. In the early 1970s an unfinished dug-out was dredged from the southern shore of Lough Neagh at Oxford Island (see Plate 10). After serious deterioration, in the absence of conservation, the vessel was sampled in 1979. This sample, QUB 3911, yielded a 228-year ring pattern which was run against the available north of Ireland chronologies. A definitive match was obtained against the Teeshan/ Drumard chronology with the outer ring at TD+14 ($t$ = 7.5). TD+14 is now known to be 492 AD and allowing for missing sapwood (the curved heartwood surface was clearly present) the dug-out was cut out of a tree felled in the range 524±9 AD. The terrible thing about this date is its extreme over-refinement. There is no need to know the date of a dug-out to this level of precision. In short, when confronted with this, the most precisely dated log-boat in the British Isles, one is forced to say, 'So what?'

At risk of being repetitive, the point is this. It is *not* that one should not try to date things like log-boats by dendrochronology. Obviously it is a relevant dating method and should be used if only because (at the moment) it is cheaper than radiocarbon. It is just that one should not expect the increased precision of date actually to improve our understanding. There are some objects which simply do not need precise dates.

## Art-historical Dating

Art-historical is a coverall term used by dendrochronologists to encompass all wooden objects related to the fine arts. The principal line of enquiry to date has revolved around works of art painted on or supported on oak panels. Studies have also been carried out on items of furniture and on statues carved out of oak.

Since the Middle Ages, artists have executed their works on wood and sheet metal as well as on canvas and paper. In the case of wooden panels the possibility exists, providing some useful species has been used, of establishing the age of the panels by straightforward dendrochronological dating. From here on the discussion relates to analyses on oak panels, as this was the main timber used for such purposes. If a painting is itself dated, then any dendrochronological estimate should help in the authentication of the picture. If, as is often the case, there is no documentary date for a painting, a dendrochronological estimate may be a valuable guide to establishing the periods of an artist's style or to separating works by the artist from works of his 'school'.

There are of course drawbacks to any study of this kind. The first relates to the problem of missing sapwood and the validity of any sapwood estimate (see Chapter 2). This factor is important because in most cases the sapwood was removed when the panels were manufactured in order to inhibit any subsequent insect attack on the painting. As a result of this precaution insect damage has not been a serious problem with panel paintings. However, from the dendrochronological point of view it results in almost all dates being estimated felling years rather than actual felling years. The second problem is akin to that discussed with regard to ships' timbers. Where did the trees grow from which the panels were split? This need not affect the dating as long as the individual ring patterns can be cross-dated against some chronology, but the question of origin should be resolved in order that the area of applicability of any resultant art-historical chronology be known.

Obviously works of art of the Middle Ages are now widely dispersed and the country in which a panel painting (or a piece of furniture for that matter) resides need not have any relation to the original country of origin. The nationality of an artist or his known area of operation will be a more important factor. However, even knowing this leaves a question over where the artist acquired his panels. So, would an artist from one country carrying out a commission in another bring with him a supply of prepared boards or could they be obtained locally? A related question must also be the length of time which would normally elapse between felling and the panel being ready for use. Was seasoning a long or short process? This might well have a bearing on whether an artist would carry his supplies with him or rely on obtaining materials at short notice. Remarkably, the workers analysing panel paintings are silent on this question, although it would seem obvious for an artist to travel equipped. Finally, with deference to the temperament of artists, were particular sources of boards 'better' than others? These questions are raised only to act as an introduction to the subject, though some further points, on the area of applicability of art-historical chronologies, are made in Chapter 13.

*Datings of Panel Paintings*

This work has to be seen principally against the background of chronology-building exercises in the northern coastal areas of Europe since the late 1960s. These chronologies were aimed at balancing the extensive central and southern German chronologies already in existence. The principal areas in this northern effort were Denmark, Schleswig-Holstein, Hamburg, Lower Saxony and, further to the west, the Netherlands (Eckstein, 1978, 117). In each area modern trees, historic and archaeological timbers were exploited to fill out the chronologies back into the first millennium AD.

It became apparent in the Netherlands that ring patterns derived from panel paintings were not compatible with ring patterns from historic buildings in that area. The panel paintings did not form an internally consistent group, though they did fall into two distinct sub-groups. Later paintings of the seventeenth and early eighteenth centuries by artists such as Rembrandt, Wouwerman and Van der Werff were on oak boards which possessed a growth characteristic consistent with Hollstein's south-west German chronology. These gave rise to Chronology I for the Netherlands, which is precisely dated both internally and by comparison with Germany. However, a second and much larger group of panel paintings from the fifteenth to the early seventeenth century yielded ring patterns of a quite different character. These panels had been used by Dutch and Flemish artists mainly in Amsterdam and Antwerp, and while they all cross-dated, the resultant Chronology II could not be definitively placed in time. As Bauch puts it, 'at present the master-chart can only be proved with a precision of ± 1 year' (Bauch, 1978a, 135). We can return to this problem later.

One of the first proving exercises was to compare the dates derived from dendrochronological analysis with known historical dates on signed paintings. This has been carried out on a relatively large number of paintings and these can be divided into those exhibiting at least some trace of sapwood and those exhibiting none. Now on the basis of the dating examples cited in Chapter 8 it might be expected that the panels with no sapwood would exhibit ring patterns spaced further back in time than those with sapwood – remember that it is not possible to tell how many heartwood rings are missing in any given case. In this respect the published results are of some interest. Bauch (1978b, 308) found that in 8 examples with sapwood, mostly paintings of Rembrandt, the average number of years between the outermost *heartwood* rings and the signed dates was 28 ± 8. This number of rings has to account not only for the sapwood but also for any seasoning or storage before use. The implication has to be that the time between felling and use was very short indeed. This finding tended to encourage faith in the 'German' sapwood estimate of 20 ± 5 years. Figure 12.5(a) shows the relationship.

Figure 12.5: (a) Relationship between outer *heartwood* rings and documentary dates for signed panel paintings which possessed some sapwood (earlier seventeenth century). (b) Frequency diagram of the intervals between final heartwood rings and documentary dates for 59 paintings earlier than the mid-seventeenth century. Compare with Figure 2.3(c). (c) Relationship between outer heartwood rings and documentary dates for paintings *c*. 1700 showing a totally different distribution.

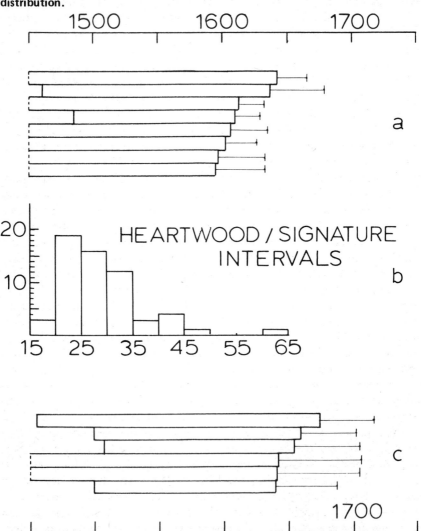

Source: Information derived from Bauch (1978b) and Fletcher (1978b).

Now when Bauch examined 15 paintings by Rembrandt and V. Goyen which had no sapwood but which had documentary dates, he found that on average there was only 26 ± 7.5 years between the outer heartwood rings and the signed dates. Given the small number of samples, these two estimates are indistinguishable and must mean that in the latter group *only* the sapwood rings have been removed and essentially none of the heartwood. This appears rational enough, since the panel-makers would presumably have attempted to maximise the width of each panel. (Sensibly the German workers do not use an actual sapwood estimate for panels without sapwood. Rather they add a minimum number of 15 rings to the outer heartwood ring and call the resultant date a 'terminus post'.) To show that this was not a unique result, Bauch also examined 24 panels from 14 paintings by Rubens. These were all part of a series ordered for Maria de Medici in 1622, none of which had sapwood. If we exclude one sample which ended in 1558, all of the others end within about 28 ± 6 years of 1622 (Bauch, 1978b, 308). All three of these estimates are broadly in line with a group of 8 paintings examined by Fletcher (1978, 305), where on average 26 ± 3.6 years elapse between the latest heartwood ring and the known date.

The outcome of these studies has to be some confidence in the relative completeness of the samples used as supports for paintings, and in the virtual absence of seasoning. (Curiously, by the early eighteenth century this tidy picture had changed. In 6 dated paintings by Van der Verff, executed between 1687 and 1716, the average time between the outer heartwood ring and signing was 51 ± 9.6 years; a totally different distribution.) However, for investigating the internal development of the style of an individual artist the lack of complete sapwood is a limiting factor. Often the only separation possible might be a suggestion of an earlier rather than a later work. For purposes of attribution and authentication the method is still very powerful, for example, where a painting is on a panel from the same tree as other paintings by the same artist. Alternatively, if a panel can be shown to have still been a living tree at the time an artist died, any painting thereon can hardly be by that artist.

Fletcher's work provides some useful examples of such exercises in resolving questions of style and documentation. One such involved a painting of the Annunciation which stylistically was thought to belong to the period *c.* 1480. On the painting was a coat of arms for an Abbot Fascet, who held office only between 1498 and 1500. Was the painting of the later period or was the coat of arms an addition? Fletcher showed that the parent tree had been alive in the early 1490s and had been felled around 1495. Clearly the stylistic dating was in error (Fletcher, 1976, 9). In another dispute Fletcher was able to show that a portrait of Jane Seymour at Woburn Abbey had been painted after her death in 1537 and most probably when her son Edward VI acceded to the throne some twenty years later.

*Netherlands Chronology II*

The question of the existence of an 'alternative' master pattern in the art-historical material comprising Netherlands Chronology II is obviously of interest to the dendrochronologist. Eckstein spells out the problem as follows: 'The curves have no marked similarity to those of Dutch mills and paintings of Chronology I, or to any of the neighbouring German chronologies.' In addition, he points out that the latest known panel of type II occurs around 1650 — thereafter only panels of type I are known — and conversely before 1650 no use was made of panels of type I (Eckstein *et al.,* 1975, 8). In a similar manner Fletcher's original MC 18 chronology, spanning 1230 to 1556, derived from art-historical material (Fletcher *et al.,* 1974), appears to show no definite agreement with any other chronologies from the British Isles (Baillie, 1978a, 30) but does match strongly with the Netherlands Chronology II.

Now what are the possibilities? These have largely been explored by Bauch (1978a, 135) and seem to be as follows. The type II trees grew somewhere, possibly in some coastal area in the Netherlands up to 1650 when they became extinct. Equally, they could all have come from some unidentified and now extinct source in England or, more implausibly, there could have been *separate,* unidentified and now extinct sources in the Netherlands and England. In the latter case these hypothetical sources — different from everything else — must have shared a strong common signal. The alternative explanation could be some unknown 'exotic' source. The arguments against imported timbers seem to centre on the fact that it is unlikely that timbers could have been continuously imported from one source for several hundred years (Bauch, 1978a, 137). Also the time between the estimated felling date and use is so short that it would leave virtually no interval for transportation.

For the interested reader, the following should be just sufficient to get the feel of the art-historical work. As with Bauch, Fletcher has over the years experienced some difficulty in tying down the English art-historical chronologies definitively. To take MC18 as an example, it was originally specified as AD 1230 to 1546 (Fletcher *et al.,* 1974). In *Nature,* volume 254 (1975, 507), this was altered to 1234 to 1550, although subsequently it was moved back to the original range (Fletcher, 1977). So by 1977 it was recognised that there were problems associated with the dating of these chronologies. The latest dating of the MC18 chronology (which formed the basis of Ref 1) as given in *Vernacular Architecture,* volume 11 (1980, 35), is 1229 to 1545, although the substantiation is not yet available.

In addition to the question of date, there are two further complicating factors with art-historical material. One relates further to the question of origin. Fletcher (1978a, 152) notes that the ring patterns of Flemish and English oak panels are very similar, but states: 'Differences in the quality of the panels and the unlikeli-hood of trade in panels on a large scale between the two countries, indicate that they were made locally.' However, a chronology constructed by Fletcher from 'Flemish' panels, namely MC14, yields a value of $t = 17.8$ when compared with the 'English' Ref 1. This is one of the highest correlation values come across in these islands and suggests that the Flemish and English ring patterns are *very similar indeed*. The second factor relates to the mean ring width of the samples. Fletcher has found it 'useful' to construct separate chronologies for narrow- and wide-ringed material. For example, Ref 1 has a mean width of 1.34 mm, while Ref 2/3 averages 1.90 mm. So, overall, while constituting an interesting and rewarding area of study, art-historical dendrochronology has given rise to a methodology rather different from most other work in the British Isles.

**Dendroclimatology**

Harking back to the Introduction, dendroclimatology is based on the hypothesis that trees growing under the same conditions within a climatic area should exhibit similar patterns of wide and narrow rings. If this hypothesis is correct, and the results presented throughout the preceding chapters seem to prove that it is, then the corollary must be that the ring widths are storing some reflection of the common signal. Since it is most likely that this signal is climatic in nature, it is not difficult to reach the conclusion that tree-rings contain, in some com-plex fashion, a record of past climate. The research based on this premiss is now well established and the interested reader is directed towards *Tree-Rings and Climate* by H.C. Fritts (1976) and to the forthcoming proceedings of the Second International Workshop on Dendroclimatology, which was held under the auspices of the Climatic Research Unit at the University of East Anglia in July 1980. The following paragraphs do no more than touch on the need for such study.

During the 1970s it has become more and more apparent that there is a need for a better understanding of past climate. We live in a world with an expanded population relying on stretched resources and not geared for fundamental changes in, for example, food supply. If changes can take place in overall climatic conditions which might affect the supplies of food, water or energy, it is im-portant to know about the likely magnitudes of such changes. Now this is not simply mindless worrying. It stems from the recognition that climatic extremes can and do occur and that their rate of occurrence and their magnitude may be well outside the limits indicated by instrumental records alone. Although some instrumental records do exist from as early as the seventeenth century, it is un-fortunately the case that, even in Europe, a really useful overall record exists only from around 1850 or later. The meteorologist in, say, the 1960s was therefore faced with the task of viewing weather in terms of the known extremes and frequencies experienced over the last century. It would be fair to say that that century had been one of few extremes. This is shown clearly in some recent observations by Lamb concerning the extremes experienced in the British Isles since 1960. There are expressed in terms of likelihood of occurrence, mostly on the basis of observations over the last century:

> in 1962-63 the coldest winter since 1740, in 1963-64 the driest winter since 1743, in 1968 and 1969 on at least four occasions 24 to 48 hour rainfalls in the lowland districts which exceeded the once in 50 years expectation, in 1974-75 the mildest winter in England since 1834, the great gale of January 2, 1976, perhaps the severest since 1703, and for 16 months to August 1976 a drought surpassing anything reported in the available raingauge records since 1727, as well as in the summer of 1976 a 24 day period warmer by about 4°C than any calendar month in the 300 year temperature record for central England (Lamb, 1977).

The occurrence of these extremes — the coldest, hottest, dryest, wettest — within such a relatively short span of time points to the inadequacy of the instrumental record as a realistic assessment of what *can* happen.

Clearly a better way of assessing extremes is to extend the data base back in time for hundreds or even thousands of years. This can be done to some extent by analysing historical information contained in diaries, documents and annals. However, these records, while giving an indication of what was happening in the distant past, are not necessarily quantifiable. They tend to reflect the experiences of individuals and this may or may not be particularly objective. As an example, take some 'extremes' noted in the compilation of early records included in the 1851 census of Ireland (HMSO, 1856). First, for 1771 we hear: 'The memory of the oldest man scarcely supplied him with an instance of so severe a season as the beginning of the year 1771.' However, only four years before (1767), in the same season, 'We began to be visited by a frost which, for severity, exceeded any that we before had experienced,' whereas the previous year (1766) 'saw the most remarkable fall of snow that has been remembered, which in some places measured more than fifteen feet'. Obviously it is not practicable to ignore these records, as they are an invaluable source of eye-witness information. They are, however, very often restricted in both area and degree of quantification.

Partly as a response to this situation with conventional records, there has been a move in recent years towards the investigation of proxy records. Certainly one of the most important sources of proxy climatic data must be tree-rings. In essence, if tree-rings contain climatic information and if this can be extracted or interpreted, it would allow the reconstruction back in time of at least some aspects of climate. Tree-rings have the particular advantage of precise time control. Thus it is possible to compare results over large areas in the same year and, since they are totally compatible with historical documentation, cross-check against specific written records. This all sounds very favourable. We want climatic information and tree-rings contain at least some. Unfortunately, the process of reconstruction is not necessarily straightforward, especially not in north-west Europe. Briefly, there are three main lines of study and these relate to the ring widths, the ring densities and isotropic compositions within the rings. The calibration of the radiocarbon time-scale, discussed above, was an obvious off-shoot of the latter.

### Reconstruction from Ring Widths

As LaMarche has so eloquently put it, 'The simplest approach to palaeoclimatic reconstruction with tree-ring data is to use a highly stratified sample from trees responding [to a first approximation] in a simple manner to a single dominant climatic variable, such as precipitation' (LaMarche, 1978, 11). To take just one example, the bristlecone pines (*Pinus longaeva* and *P. aristata*) in the White Mountains of California show a sharp cut-off at the upper limit of their distribution — the upper tree-line. That conditions have varied in the past is clearly shown by the remains of dead trees at altitudes up to 150 metres above the present-day tree-line. LaMarche was able to argue that the mean annual ring width at the upper tree-line was an approximate record of warm-season temperatures in the White Mountains. By simply meaning the raw ring widths (not indices) over the last 5,500 years and plotting the mean figure of each 100-year block, he established a picture of warm-season temperature which tied in adequately with other evidence (LaMarche, 1974, 1046).

An extension of this 'simple' approach is possible where large numbers of chronologies are available in a region. This involves the construction of relative departures of ring indices for blocks of years across a regional array of data points. In effect, this allows the plotting of tree-growth contours in a spatial pattern. Modern pattern can be calibrated against known atmospheric pressure anomalies, allowing the interpretation of similar patterns in the past. Due to the large number of available chronologies in the western United States, it was obvious that such work would be pioneered there (Fritts, 1965). Hopefully the growing number of modern chronologies in the British Isles (see Chapter 4) and Europe may allow similar work in this region in the near future.

###  X-ray Densitometry and Isotropic Ratios

If we accept that ring widths reflect at least some aspects of climate, then it is not unreasonable that the relative compositions of the rings may be recording changes in aspects of their environments. X-ray densitometry involves the measurement of the density profiles across each ring. Obviously in the same way that cell size and wall thickness alter across the ring, the density also alters. This gives a profile for each ring which is low for the early wood and high for the late wood. Since there are difficulties in making objective interpretations of the overall profile, the study is normally simplified to measurement of the maximum density (late-wood density) of each ring. The result of densitometry measurement on a timber sample is therefore a set of numbers not unlike a ring pattern.

One of the main centres for such research in Europe is the Swiss Federal Institute of Forestry research at Birmensdorf. Workers there have shown that the maximum density of rings in conifers from sub-alpine, cool, humid regions is primarily controlled by summer temperature. In contrast, trees on dry sites have densities primarily associated with summer precipitation. By analysing trees from both types of site it is possible to characterise the summer temperature and rainfall for each year, as far back as the tree-ring chronologies extend. Rather in the same way that spatial contours of ring-width departures suggested pressure anomalies in the western United States, the Swiss workers observe spatial patterns of maximum density. There spatial patterns suggest identical growth behaviour in cell-wall construction in pine and spruce trees in particular years over areas extending from the Alps to Scotland to southern Scandinavia (Schweingruber *et al.*, 1979). While this very spatial consistency demonstrates the repeatable character of maximum density measurements, the progress with the interpretation of oxygen isotope ratios in tree-rings has faltered due to the inability of different workers to duplicate results.

In principle the isotope method is as follows. The ratios of the stable isotopes of hydrogen and oxygen vary with the air temperature prevailing when the ring was formed. Thus each ring should contain an isotope temperature record. Exactly the same phenomenon is known to occur with the uptake of carbon in plants. The ratios of the isotopes $^{12}$C, $^{13}$C and $^{14}$C vary from sample to sample. The recognition of the possible effect of this variation is implicit in the fractionation correction applied to each radiocarbon date. There are problems with isotope ratios, however, not the least of which is the difficult chemistry involved in extracting oxygen from wood cellulose. One impressive set of results on the $^{18}$O ratio in German oaks is worthy of mention as it appears to show extremely good agreement between the $^{18}$O ratios and annual average temperature over the period 1700 to the present (Libby *et al.*, 1976, 284).

It seems likely that between the three methods, widths, densities and isotopic ratios, notwithstanding other approaches which may become available, it should be possible to make adequate reconstructions for the last thousand years with thinner but none the less useful extrapolations well back into the post-glacial.

# Chapter 13

# Future Developments

One of the rather attractive aspects of dendrochronology is the fact that once a reference chronology is completed, and specified exactly in time, it can be used to date timbers not only now but at any time in the future. For example, the chronology constructed back to AD 700 by Douglass in 1929 is just as relevant today as it was then. Chronologies relate to past events, the growth of trees under changing year-to-year conditions. The resultant signal, common to trees for any given period within an area, is fixed. Once that signal has been reconstructed it is relevant for all time. So, as long as archaeologists continue to uncover timbers, it will be possible to provide accurate dates. The future for dendrochronology as a dating method seems assured.

It is intended to review here the main periods of relevance and their likely development. Before doing so it is necessary at least to touch upon one aspect of dendrochronology in these islands which has to be fundamental to our understanding of the medium with which we are dealing. In broad terms this relates to the number of tree-ring areas within the British Isles. In 1973 the present author was still restricting all sample collecting to the area constituted by the north of Ireland. Figure 4.2 shows this area in relation to the British Isles as a whole (work on the Dublin timbers was carried out in isolation). Since there was very little to go on, it would not have been surprising to discover that the British Isles was made up of a large number of small, discrete and possibly mutually exclusive tree-ring areas. However, as I hope has become clear in the preceding chapters, the position was relaxed as more and more chronologies for different areas within the British Isles, and for a variety of time periods, were constructed. The consistent good cross-agreements encountered prompted the idea that there might be an Irish Sea basin tree-ring area (Baillie, 1978a) with possible peripheral areas which might or might not be separate entities, for example south-western Ireland and south-eastern England. One chronology stood out, at the time, as being different from the others in Britain and Ireland. This was MC18 (Fletcher *et al.*, 1974), an art-historical chronology constructed mostly from the oak boards of panel paintings, which showed no significant agreement with any other British Isles chronologies.

As a response to this difference and to the Irish Sea basin idea, Fletcher proposed that in the British Isles there are two types of chronology for the Middle Ages — broadly the period 1250 to 1600. His type A (art-historical) relates to MC18 and subsequent variants Refs 1-5. All others belong to variants of type H (Huber type). This terminology relating to Huber was based on the observation that some English chronologies cross-dated directly with the central German chronologies. The Belfast chronology was assigned to type H because of the stepwise agreement from Ireland to the England/Wales (Giertz) chronology to Germany. The two types A and H appear to be mutually exclusive. Since the English type A chronologies showed good cross-agreement with the art-historical, Netherlands type II chronology (see Chapter 12), Fletcher proposed a North Sea coastal area to balance the Irish Sea basin proposal, viz. 'Type A may be as wide-spread, embracing lowland coastal areas around the whole southern basin of the North Sea' (Fletcher, 1978, 154).

Now type H shows up consistently throughout the whole of the last two millennia. In Chapters 6, 7 and 9 we saw consistent stepwise correlations between all the Belfast-derived chronologies and Germany. Type A, namely chronologies which do not cross-match with type H, so far only occur for art-historical and related timbers. This observation prompted the author to suggest that the art-historical material was all imported (Baillie, 1978a). This was refuted by Fletcher on the grounds that too many diverse objects were of type A for them all to be imports. In order to explain the non-compatibility of the type A chronologies, if they did in fact derive from English timbers, Fletcher suggested the following hypothesis: since the type A chronologies were derived from the south and east of England, there were two factors which separate the 'east' from the 'west' in the British Isles. One was rainfall; the dotted line in Figure 4.2 shows the 30 inch/annum limit used by Fletcher to divide the British Isles into two zones. The other factor was terrain. The east is basically low-lying, while the west is hilly (more like the area in Germany from which Huber derived his chronologies). A third factor could then logically be added, since sessile oaks are found predominantly in the hilly areas of the west and pedunculate in the east. This division is in no way absolute, simply a tendency.

So by 1978 there were clearly two schools of thought developing within the British Isles — Baillie and the idea of uniformity on the one hand, and Fletcher with the idea of differing responses on the other. Unfortunately there is no easy answer to the questions posed, but some thoughts are offered to try to delimit the problem.

If between 1250 and 1600 there were trees growing in England which were genuinely *different*, namely type A (since type H have always been present we need not argue about them), this implies that conditions in the past were different. However, Fletcher's east/west, dry/wet, lowland/hilly and pedunculate/sessile

divisions are based on modern differences. Unfortunately for that line of argument there are no traces of type A chronologies at the present time. The results
outlined in Chapter 4 were aimed partly at resolving this question. It seems clear
that for the recent past trees in the British Isles are responding to a common
signal. There is no site chronology outstanding in its difference — the uniformity
is quite remarkable. However, in Chapter 4 the question of species was not
entered into, except to disregard it. All the chronologies were used without
regard to species, though it was suspected that a majority might be sessile. An
independent test of the species theory is possible. In 1974 Fletcher had published
two chronologies from Bagley Wood, Berkshire. This site was within the low-
rainfall area and had yielded a chronology for each species. How would these two
species chronologies relate to the overall British Isles 18-site master chronology?
Table 13.1 shows the result of comparing each with the overall master (these
two chronologies are not included in the master and therefore cannot bias it in
any way).

**Table 13.1: Comparison of Two Species Chronologies with the Overall British Isles Master Chronology**

|          | No. of Trees | Length    | $t$ cf BI 18 | Species            |
|----------|--------------|-----------|--------------|--------------------|
| Bagley 1 | 10           | 1836-1971 | 5.2          | mostly pedunculate |
| Bagley 2 | 4            | 1881-1968 | 4.8          | sessile            |

Clearly both of these chronologies contain a strong element of the overall signal.
So there is as yet no sign of a modern type A chronology on the basis of species,
on the basis of altitude or on the basis of dryness (Norwich and Oxford chronologies, as well as Bagley, lie within the dry eastern area: see Figure 4.4).

Where does the solution lie? There are really only two possibilities. Either
conditions were very different in the 1250 to 1600 period, so different as to
allow the growth of mutually exclusive type A and H patterns, or the type A
chronologies are based on exotic timbers.

## The Main Periods of Interest

What follows is a brief resumé of the main periods of interest from a dendrochronological point of view. For recent centuries the main stimulus is likely to
be the growing interest in dendroclimatology cited in Chapters 4 and 12. It is

likely that in the next few years we should see the construction of sufficient living-tree chronologies within the British Isles to gain an almost complete understanding of the areal variations within these islands. Not only will it resolve one way or the other the question of type A and type H chronologies for the modern period, but it may well force an extension of our interest into the growth of other species. This is something which has already had extensive coverage on the Continent. In fact, taken in conjunction with the parallel expansion of chronologies throughout Europe, the next decade may well see a very refined picture of climatic variation emerging.

The results presented in Chapters 4 to 11 showed that we cannot treat dendrochronology as a continuous study. The next major unit of 'tree-ring time' runs from the seventeenth to the fourteenth centuries. In this respect the English Refs 1 to 5 are also different, as they do not appear to show the otherwise ubiquitous fourteenth-century gap. Leaving them aside, the picture for the remainder of the later medieval chronologies is relatively coherent. The chronologies for south-west Scotland, northern Ireland, the Dublin area, the north-west of England (Hughes, personal communication), Sheffield, the north-east of England, the Welsh borders and the south of England[1] (Fletcher, personal communication) all cross-match with most, if not all, the others. Any fresh chronologies for this period should cross-match with at least several of these.

The only areas about which any question remains are north and east Scotland, the south and west of Ireland and the extreme east of England. It is unlikely that any serious effort will be made in the near future to tackle the Irish south-west, simply because there are so few buildings with timbers that the results would not justify the effort involved. If the area is ever to have a medieval chronology, the stimulus may again come from climatologists interested in extending the proxy data base back in time. The most pressing need would appear to be the construction of a chronology for East Anglia, first to fill out the picture, and, second, to date some of the host of vernacular buildings in the area. In Scotland most of the timbers from this period are from the extreme south-west. Some effort should probably be made to pull together a chronology from the north and east, drawing on timbers from Perth, Fife and sources further to the north. However, there is little point in doing this unless there are or are likely to be some specific chronological questions to be answered.

The next period is roughly defined as the ninth- to fourteenth-century medieval unit. In some ways this is the archaeological period with the best chronology cover at the present time. With chronologies for Perth, Glasgow, Dumfries, Belfast, Dublin, Nantwich (Liverpool area), Exeter, London and York covering some or all of this period, there can be few sites anywhere in the British Isles which cannot be cross-dated. Moreover, for this period most of the chronologies are highly correlated and those in the south correlate directly with the established German chronologies. This represents a highly integrated picture and one with important implications for medieval archaeologists who can anticipate routine dating of most oak samples.

The chronologies discussed in Chapter 9, namely those for the northern and southern halves of Ireland and Tamworth, combined with Fletcher's Ref 8, mark the beginning of a grid of Dark Age chronologies. Hillam already has at least two other central English site chronologies of this period. The one area which stands out is Scotland and there the situation is fairly serious. There seem to be very few available Dark Age timbers in Scotland. The few pieces come across by the author, namely one sample from Dundurn and one from Loch Maben, suffer from short ring patterns and isolation. The shortage of good groups of timbers is probably a reflection of the small number of excavations on the types of site likely to provide timbers. Short of some major accidental find of suitable timbers, it is unlikely that any serious progress will be made with this period in Scotland. If on the other hand there are specific problems to be solved in the first millennium in Scotland, it will be increasingly up to the archaeologists to direct their efforts towards finding the relevant timbers.

**Earlier Periods**

It is ironic that Ireland should have a completed chronology across the whole of the Roman period and no Roman sites. In fact there are very few sites of the first half of the first millennium in Ireland, wood-bearing or otherwise (known to archaeologists). However, in England it is only a matter of time before one, or several, chronologies of this period become available. These chronologies will almost certainly be tied down by cross-dating against either the Belfast or the Hollstein/Becker chronologies, preferably both. In England there is already a 282-year floating chronology derived from Roman waterfront timbers (Morgan and Schofield, 1978). This chronology is of interest because on archaeological grounds it is thought to belong to the second century AD. If so, it would span roughly the period 100 BC to 200 AD. However, there are four radiocarbon dates associated with the chronology which in the author's opinion would be *much* more at home with timbers of the very late fourth century. It will be interesting to see if the radiocarbon dates turn out to be correct. If they do, we will be seeing another example of the problem encountered by Hollstein (see the relative dating section in Chapter 12).

For the prehistoric period the author has already made his views clear. There are strictly limited applications in the British Isles for prehistoric dendrochronology. Too often individuals think that to build such a chronology would be a 'nice thing to do' and do not think through to the logical conclusion, which is

'What are we going to do with it when we've got it?' Alas, it may not transpire until too late that the applications are few and far between. There is still, as stressed before, too much emphasis on the novelty value of dendrochronology. The time is approaching when the archaeological 'public' will no longer be impressed with the dating of known-age buildings or irrelevant scraps of timber! Having said that, there are some heartening words for the serious student of chronology. While there are still some problems to be solved, it is reasonable to suggest that the overall picture of dendrochronology in the British Isles is beginning to emerge. As time goes on, more and more of the essential reference chronologies are becoming available. Broadly speaking, there is adequate coverage for the last millennium. The situation for the first millennium is not far behind and things look hopeful for the dating of a range of early medieval remains. The Roman period has got to be sorted out in the very near future.

With regard to the Belfast long chronology, certainly the nearest to completion in Europe, it is interesting to see how things have changed. In the original draft of this book the author was suggesting that the long chronology would never be completed. Now with the steady refinement of the high-precision radiocarbon work and with the better understanding of the Suess calibration, the prospect of region-to-region bridging (Ireland to England to Germany) holds at least some hope of one 7,500-year European chronology.

## Note

1. This south of England chronology, for the middle ages, is one derived from building timbers. Unlike the art-historical chronologies REF 1-5 which do not match with other British Isles chronologies, this chronology shows consistent agreement and hence can be assumed to be indigenous.

# Crannogs

These archaeological structures are essentially artificial islands constructed for the purpose of living in the protected isolation of a body of water. They tend to be rather loosely defined partly because, in the absence of extensive excavations, it is difficult to specify in any individual case just how artificial they are. In many instances, therefore, we may be looking at small islands which were merely consolidated by man, as opposed to totally artificial structures. From what we do know (and from a certain amount of personal observation) it is possible to suggest the sort of range of structures encompassed by the term crannog.

The body of the crannog can be built up of stone, timber, brushwood or unspecified organic material or, of course, any combination of these. The decision in any individual case was almost certainly conditioned by which materials were available. However, within the structures which interest the dendrochronologist, those containing substantial oak timbers, there seem to be two clear groups. The first represents sites which were built up by random (?) dumping of tree trunks, for example the Fermanagh crannogs. The second group appears to represent a more systematic approach where a definite framework of riven-oak beams was constructed and used as a basis for construction. The crannog at Teeshan, Co. Antrim, was apparently one of these (personal communication from Mr R. Warner, who witnessed the destruction).

So if we ignore improved natural islands, which seem to be a trivial case there were two ways to build a crannog. The simplest would be to find a natural shallow or shoal. A ring of stakes would then be pushed into the lake-bed and hundreds of journeys made by dug-out, carrying materials to raise up a platform above water level. Now you may ask how the builders would know that any given level would be adequate, i.e. that the whole thing would not be submerged in the next wet spell? It seems likely that they would build in the winter for three reasons: (a) because the water level would be at its highest and hence they could dump directly on to the higher parts of the artificial island (an important consideration when moving large boulders in an open boat); (b) they could make use of frozen conditions to slide the material out in at least some seasons (albeit an idea which would seem to negate the whole point of having a crannog!);

(c) the winter might be the period when most labour was available for the construction.

The second method would seem to apply best to construction in deeper water. Here a prepared wooden framework would be towed out (?) and then weighted and sunk in position. The crannog would then be constructed as before with the proviso that a linked ring of timbers be available to use as a coffer (to hold the material together). The sites in the text which yielded dates were a mixture of both types. Interestingly, Teeshan, with a framework, had timbers of two dates. This could be interpreted as the use of already available timbers to make the framework. The other dates were mostly for random trunks of trees which would presumably constitute immediate use.

The one thing which becomes apparent is the very considerable labour involved in the construction of even a modest crannog. Considering that the end product, assuming you built a house on top, was a damp and probably revolting home, the stimulus behind their construction must have been pressing.

# Horizontal Mills

One point needs to be made clear. The 'horizontal' mill in Ireland, and hopefully in Britain, relates to the type of device shown in Figure 9.2. The wheel lies in the horizontal plane. The vertical shaft drives the top stone of the mill directly. This terminology goes back at least to the mid-nineteenth century (McAdam, 1856) and hence it is surprising to find such mills being described as *vertical* (because the shaft is vertical) as late as 1975 (Horn, 1975, 226). It should be obvious that any description based on the shaft is ambiguous, since in all mills the shaft must be vertical *at some point*.

Remains of something like 70 horizontal mills are known to have been discovered in Ireland in the last two centuries. By combining the various remains from different sites it is possible to get a reasonably complete picture of what one of these structures was like. The mill consisted of a wheel-house, normally dug into the bed of a stream or set in a low-lying spot to which a stream or spring could be diverted. Dimensions vary, but a number seem to have been about 2m wide by 3m in length. The wheel-house normally survives as two very heavy side beams (halves of a single tree) and a cross piece at the upstream end. In some instances the wheel-house had a plank floor. The total height of this under-house is indicated by a complete wheel and shaft from Moycraig, Co. Antrim (McAdam 1856) now in the Ulster Museum. This unit is almost exactly 2m in length.

In practice, the shaft would have risen through a floor which supported the millstones. There is some evidence that the whole structure would have had a thatched roof (Baillie, 1975). In operation, water was jetted through the narrowing flume (a tapering trough as in Plate 8) into the scoop-shaped paddles on the wheel. In instances where there was a poor water supply the flume may have been almost vertical, to 'drop' the water into the paddles. The wheel itself was a complicated piece of carpentry with about 20 paddles tenoned into the hub and held by wooden pegs. As the water rotated the wheel, so it rotated the top stone which was keyed on to an iron (?) sile on the end of the shaft (see Figure 9.2). In order to vary the 'grind' of the mill, the whole rigid wheel-shaft-top-stone assembly was seated on a longitudinal plank, the sole tree, which was capable of

259

a small amount of vertical movement via a lever system.

To stop the mill it is assumed that a system of sluices was available to divert the water. All in all, these mills represent a simple but elegant technology clearly in use in Ireland for at least three hundred years, from 630 to 930 AD. Given the random nature of the sampling for dendrochronology, it must be doubtful if any of the archaeologically discovered examples fall outside this range — certainly the majority must fall within it. If, as is suspected, these mills are to some extent the by-product of monasticism, one still has to ask what happened after the latest ones were constructed around 930. Is it simply that they were replaced by the vertical mill of which we find so little evidence?

# What the Archaeologist Needs to Know

Surprisingly, this is not a long section. The plain fact of the matter is that dendro-chronology is only ever difficult to the dendrochronologist. To the archaeologist who supplies samples for dating there are only three possible outcomes. Either he gets dates or he does not (the third possibility is that the dates are wrong). Let us therefore divide the problem into two. On the one hand, there is the archaeologist and his expectations, on the other the dendrochronologist and his 'real life' dating method. We will look at how they compare (see also Hillam, 1979).

## Expectation

Archaeologists assume that dendrochronologists can date any timber species, from any area and from any time period, no matter how small or rotten the sample.

### *Real Life*

Dendrochronologists will specify oak (in the British Isles) as the only timber which is realistically datable, with provisos about elm and ash under certain circumstances. Short-lived samples can seldom be reliably dated.

### *Advice*

Give your dendrochronologist long-lived oak samples of broadly the last two millennia and you stand some chance of getting dates. Better still, wherever possible provide groups of timbers as the resultant master chronology stands a better chance of dating, and there is the possible bonus of relative information on phases, re-use, etc.

## Expectation

Dendrochronology gives precise dates.

*Real Life*

If you want precision, then you must supply complete samples, i.e. samples with sapwood. Incomplete samples can still give better dates than probably any other method, but they will not give precise dates.

*Advice*

Check with your dendrochronologist (a) the dating quality and (b) any sapwood estimate used. Where possible, let the dendrochronologist carry out his own sampling. After all, he has to do the dating.

## Expectation

All tree-ring samples should be kept wet.

*Real Life*

With excavated samples, especially those with sapwood, this is probably a good idea. Samples allowed to dry out tend to lose sapwood.

*Advice*

If samples have dried out, they are still perfectly useful for dating purposes, so keep them. If the wood has no sapwood and seems sound, slow air drying will do no harm and makes the samples a lot easier to handle. Finally, if you need precise dates, save the sapwood at all costs. Remember it is soft and easily damaged — especially by the dense heartwood to which it is attached.

## Expectation

All dendrochronologists are equal.

*Real Life*

Tricky! If dendrochronology is treated as a science, using visual and statistical cross-matching backed up by replication, then the dates should be repeatable, i.e. any laboratory should give the same result.

*Advice*

Find out if the date is backed up by a statistical correlation value — even better if it matches with more than one chronology at the same year.

## Expectation

Archaeologist quote: 'If it doesn't give a definite date, can you [the dendro-chronologist] suggest some possibilities or give a rough date?'

### Real Life

Remember that a ring pattern either dates or it does not. There should be no half-measures. If the ring pattern ties up with a *floating* chronology, then you should get a floating date.

### Advice

Do not accept multiple-choice dates. This is a particularly dangerous game, willingly entered into by the archaeologist if given a chance. You will pick the possible date which suits your own ideas and use it to reinforce those ideas. Hence a circular argument – worse still a reinforced circular argument. What the dendrochronologist is doing in such a case is listing the highest correlation values without being able to specify a definite match. In fact, none of them may be correct! To put it bluntly, if a dendrochronologist tells you that a timber is *probably* of a certain date – change your dendrochronologist.

## Expectation

'If I keep pressuring him [the dendrochronologist] he's sure to produce a date.'

### Real Life

A tree-ring sample either dates or it does not. No amount of pressure will *make* a ring pattern match if it does not. Similarly, if you submit a sample from the fossil forest, you may have to wait some time for your date.

### Advice

A touch of hyperbole. Imagine you are a subordinate junior (of a dendrochron-ologist) and in waltzes the Regius Professor wanting to know why you *still* haven't dated his mongongo root netsuke – after all, you've had it a week. 'Well, grovel mumble grovel,' you say, 'it's not quite dated yet, your omnipotence.' 'What,' barks the professor, 'you mean it isn't dated *yet*?' 'Well actually,' you say, 'there is a possible tie up with a ... uhm ... [quick thinking] ... fragment of the Original ... no ... oh yes, with a Maori club from Cook's first voyage, sir.' 'Great,' says the professor, and goes off happily with his eighteenth-century – 'tree-ring dated, you know' – netsuke.

Well, what would you have done in the same position? (But in truth, even though you may never work again, you should have told him what to do with his mongongo root!)

# References

Note: A large number of references are to papers in 'Dendrochronology in Europe' (ed. J.M. Fletcher), *British Archaeological Reports,* International series, 51. This is abbreviated throughout to *BAR* International series, 51.

Alcock, L. (1977) 'Excavations at Dundurn, St. Fillans, Perthshire 1976-7' interim report, University of Glasgow

Apsimon, A.M. (1976) 'Ballynagilly and the Beginning and End of the Irish Neolithic' in S.J. De Laet (ed.), *Acculturation and Continuity in Atlantic Europe,* Ghent, pp. 15-30

Atkinson, E.D. (1934) *An Ulster Parish; Waringstown,* Dundalk

Baatz, D. (1977) 'Bemerkungen zur Jahrringchronologie der Romischen Zeit', *Germania, 55,* 173-9

Baillie, M.G.L. (1973a) 'A Dendrochronological Study in Ireland', PhD dissertation, Queen's University, Belfast

—— (1973b) 'A Recently Developed Irish Tree-ring Chronology', *Tree-Ring Bulletin, 33,* 15-28

—— (1974a) 'Dendrochronological Dating of the Oak Timbers from the Gloverstown House', *Ulster Folklife, 20,* 41-5

—— (1974b) 'A Tree-ring Chronology for the Dating of Irish Post-medieval Timbers', *Ulster Folklife, 20,* 1-23

—— (1975) 'A Horizontal Mill of the Eighth Century AD at Drumard, Co. Londonderry', *Ulster J. Archaeol, 38,* 25-32

—— (1976) 'Dendrochronology as a Tool for the Dating of Vernacular Buildings in the North of Ireland', *Vernacular Architecture, 7,* 3-10

—— (1977a) 'An Oak Chronology for South Central Scotland', *Tree-Ring Bulletin, 37,* 33-44

—— (1977b) 'The Belfast Oak Chronology to AD 1001', *Tree-Ring Bulletin, 37,* 1-12

—— (1977c) 'Dublin Medieval Dendrochronology', *Tree-Ring Bulletin, 37,* 13-20

—— (1978a) 'Dendrochronology for the Irish Sea Province' in P. Davey (ed.), *Man and Environment in the Isle of Man BAR,* British series, *51,* 27-37

—— (1978b) 'Dating of Some Ships' Timbers from Wood Quay Dublin', *BAR,* International series, *51,* 259-62

—— (1978c) 'Scottish Dendrochronology', *Scottish Forestry, 32,* 255-8

—— (1979a) 'An Interim Statement on Dendrochronology at Belfast', *Ulster J. Archaeol, 42,* (forthcoming)

—— (1979b) 'Some Observations on Gaps in Tree-ring Chronologies, *Proc. Symposium on Archaeol. Sciences* (Jan. 1978), University of Bradford, pp. 19-32

—— (1980) 'Dendrochronology; the Irish View', *Current Archaeol. 7,* (2), 61-3

—— (1981a) 'Perth High Street Timbers; the Dating', *Report on the Perth Excavations* (forthcoming)

—— (1981b) 'Dendrochronology' in M.J. O'Kelly, *Festschrift,* Cork (forthcoming)

—— and Pilcher, J.R. (1973) 'A Simple Cross-dating Program for Tree-ring Research', *Tree-Ring Bulletin, 33,* 7-14

Bannister, B., and Robinson, W.J. (1975) 'Tree-ring Dating and Archaeology', *World Archaeol, 7,* (2), 210-25

Barefoot, A.C. (1975) 'A Winchester Dendrochronology for 1635-1972 AD, its Validity and Possible Extension', *J. Inst. Wood Science, 7,* 25-32

——, Hafley, W.L., and Hughes, J.F. (1978) 'Dendrochronology and the Winchester Excavation', *BAR* International series, *51,* 162-71

Barry, J. (1962) *Hillsborough; a Parish in the Ulster Plantation,* Belfast

Basch, L. (1972) 'Ancient Wrecks and the Archaeology of Ships', *Int. J. Nautical Archaeol., 1,* 1-58

Bass, G. (1972) 'The Earliest Seafarers in the Mediterranean and Near East' in G. Bass (ed.), *A History of Seafaring,* London

Bauch, J. (1978a) 'Tree-ring Chronologies for the Netherlands', *BAR* International series, *51,* 133-7

—— (1978b) 'Dating of Panel Paintings', *BAR* International series, *51,* 307-14

Becker, B., and Delorme, A. (1978) 'Oak Chronologies for Central Europe. The Extension from Medieval to Prehistoric Times', *BAR* International series, *51.* 59-64

—— and Schirmer, W. (1977) 'Palaeoecological Study on the Holocene Valley Development of the River Main, Southern Germany', *Boreas, 6,* 303-21

Beckett, J.C. (1944) *The Making of Modern Ireland,* London

Bentley, ?. (1861) *The Life and Letters of Mary Granville,* London

Bersu, G. (1947) 'The Rath in Townland Lissue, Co. Antrim', *Ulster J. Archaeol., 10,* 30-58

Brett, D.W. (1978) 'Medieval and Recent Elms in London', *BAR* International series, *51,* 195-9

Camblin, G. (1951) *The Town in Ulster,* Belfast

Campbell, J.A., Baxter, M.S., and Alcock, L. (1979) 'Radiocarbon Dates for the Cadbury Massacre', *Antiquity, 53,* 31-8

Clarke, E. (1967) 'The Colonization of Ulster and the Rebellion of 1641' in T.W. Moody and F.X. Martin (eds.), *The Course of Irish History,* Cork

Cook, O. (1968) *The English House Through Seven Centuries,* London

DeBreffny, B., and Mott, B. (1977) *Castles of Ireland,* London

De Jong, A.F.M., Mook, W.G., and Becker, B. (1979) 'Confirmation of the Suess Wiggles 3200-3700 BC', *Nature, 280,* 5717, 48-9

Delaney, T.G. (1974) 'Summary Account of Excavations in Carrickfergus' in T.G. Delaney (ed.), *Excavations,* 7

—— (1976) 'Summary Account of Excavations in Carrickfergus' in T.G. Delaney (ed.), *Excavations,* 4-5

Delorme, A. (1972) 'Dendrochronologische Untersuchungen an Eichen des Sudlichen Weser- und Leineberglandes', doctoral dissertation, University of Gottingen

Dolley, M., and Seaby, W.A. (1971) 'A Find of Thirteenth Century Pewter Tokens from the National Museum Excavations at Winetavern Street Dublin', *Spink's Numismatic Circular* (Dec.), 446-8

Douglass, A.E. (1919) *Climatic Cycles and Tree Growth I,* Washington

—— (1928) *Climatic Cycles and Tree Growth II,* Washington

—— (1937) 'Tree Ring Work; 1937', *Tree-Ring Bulletin, 4,* 3-6

Eckstein, D. (1978) 'Regional Tree-ring Chronologies along Parts of the North Sea Coast', *BAR* International series, *51*, 117-24

—— and Bauch, J. (1969) 'Beitrag zu Rationalisierung eines dendrochronologischen Verfahrens und zu Analyse siener Aussagesicherheit', *Forstwis Centralbl., 88,* 230-50

——, Brongers, J.A., and Bauch, J. (1975) 'Tree-ring Research in the Netherlands', *Tree-Ring Bulletin, 35,* 1-13

——, Mathieu, K., and Bauch, J. (1972) 'Jahrringanalyse und Baugeschichtsforschung, aufbau einer Jahrringchronologie für die Vier- und Marschlande bei Hamburg', *Abhandlungen und Verhandlungen des naturwiss. Vereins Hamburg, 16,* 73-100

Esau, K. (1960) *Plant Anatomy,* New York

Evans, E.E. (1957) 'The Ulster Farmhouse', *Ulster Folklife, 1,* 27-31

Fahy, E.M. (1956) 'A Horizontal Mill at Mashanaglass', *J. Cork Hist. Archaeol. Soc., 61,* 13-57

Farrell, A., and Baillie, M.G.L. (1976) 'The Use of Dendrochronology in Nautical Archaeology', *Irish Archaeol. Research Forum, 3* (2), 45-55

Ferguson, C.W. (1968) 'Bristlecone Pine; Science and Esthetics', *Science, 159,* 839-46

—— (1969) 'A 7104 Year Annual Tree-ring Chronology for Bristlecone Pine (*Pinus aristata*) for the White Mountains of California', *Tree-Ring Bulletin, 29* (3-4), 3-29

Fletcher, J.M. (1974) 'Annual Rings in Modern and Medieval Times' in M.G. Morris and F.H. Perring (eds.), *The British Oak,* Faringdon, pp. 80-97

—— (1976) 'Oak Antiques; Tree-ring Analysis', *Antique Collecting and Antique Finder* (Oct.), 9-13

—— (1977) 'Tree-ring Chronologies for the 6th to 16th Centuries for Oaks of Southern and Eastern England', *J. Archaeol. Science, 4,* 335-52

—— (1978a) 'Oak Chronologies; England', *BAR,* International series, *51,* 145-56

—— (1978b) 'Tree-ring Analysis of Panel Paintings', *BAR,* International series, *51,* 303-6

——, Tapper, M.C., and Walker, F.S. (1974) 'Dendrochronology; a Reference Curve for Slow Grown Oaks AD 1230 to 1546', *Archaeometry, 16,* 31-40

Fritts, H.C. (1965) 'Tree-ring Evidence for Climatic Changes in Western North America', *Monthly Weather Review, 93,* 421-43

—— (1976) *Tree-Rings and Climate,* London

Gailey, A. (1974) 'A House from Gloverstown, Lismacloskey, Co. Antrim', *Ulster Folklife, 20,* 24-41

—— and McCourt, D. (1978) 'A List of North Irish Crucks', *Vernacular Architecture, 9,* 3-9

Graybill, D.A. (1979) 'Revised Computer Programs for Tree-ring Research', *Tree-Ring Bulletin, 39,* 77-82

Hall, E.T. (1946) 'Preserving and Surfacing Rotted Wood and Charcoal', *Tree-Ring Bulletin, 12,* 26-7

Hatcher, J. (1977) *Plague, Population and the English Economy 1348-1530,* London

Haury, E.W. (1962) 'HH-39; Recollections of a Dramatic Moment in Southwestern Archaeology', *Tree-Ring Bulletin, 24,* 11-14

Heyworth, A. (1978) 'Submerged Forests around the British Isles', *BAR* International series, *51,* 279-88

Hillam, J. (1976) 'The Dating of Cullyhanna Hunting Lodge', *Irish Archaeol. Research Forum, 3* (1), 17-20

—— (1979) 'Tree-rings and Archaeology; Some Problems Explained', *J. Archaeol. Science, 6,* 271-8

HMSO (1856) *The Census of Ireland for the Year 1851,* Part 5: Tables of Deaths, vol. I,

Dublin

HMSO (1940) *Preliminary Survey of Ancient Monuments for Northern Ireland*, Belfast

Hollstein, E. (1965) 'Jahrringchronologische von Eichenholzern ohne Waldkande', *Bonner Jahrb., 165*, 12-27

—— (1979) 'Mitteleuropaische Eichenchronologie', Mainz am Rhein

Horn, W. (1975) 'Water Power and the Plan of St Gall', *J. Medieval Hist., 1*, 219-57

Huber, B. (1967) 'Dendrochronologie', *Acta Bernensia*, 145-56

—— and Giertz, V. (1969) 'Our 1000 Year Oak Chronology', *Conference Report to the Austrian Academy of Science, 178*, 32-42

—— and Giertz, V. (1970) 'Central European Dendrochronology for the Middle Ages' in R. Berger (ed.), *Scientific Methods in Mediaeval Archaeology*, Berkeley and Los Angeles

Hutchinson, W.R. (1951) *Tyrone Precinct*, Belfast

Jones, E.W. (1959) '"Quercus L." Biological Flora of the British Isles', *J. Ecol., 47*, 169-222

Kapteyn, J.C. (1914) 'Tree Growth and Meteorological Factors', *Recueils Trav. bot. Neerland, 11*

Kolchin, B.A. (1962). 'Dendrochronology of Novgorod', *Soviet Archaeol., 1*, 113-39 (see also Thompson, M.W.)

LaMarche, V.C. Jr. (1974) 'Palaeoclimatic Inferences from Long Tree-ring Records', *Science, 183*, 1043-8

—— (1978) 'Tree-ring Evidence of Past Climatic Variability', *Nature Climatology Supplement* (23 Nov.), 8-12

—— and Harlan, T.P. (1973) 'Accuracy of Tree-ring Dating of Bristlecone Pine for Calibration of the Radiocarbon Timescale', *J. Geophys. Research, 78* (36), 8849-59

Lamb, H.H. (1977) 'Climatology I', *Times Higher Education Supplement* (21 Jan.)

Laxton, R.R., Litton, C.D., Simpson, W.G., and Whitley, P.J. (1979) 'Dendrochronology in the East Midlands', *Trans. Thoroton Soc. Nottinghamshire*, 23-4

Lees-Milne, J. (1970) *English Country Houses; Baroque 1685-1715*, Norwich

Leggett, P., Hughes, M.K., and Hibbert, F.A. (1978) 'A Modern Oak Chronology from North Wales and its Interpretation', *BAR*, International series, *51*, 187-94

Lenox-Conyngham, M. (1947) *An Old Ulster House; Springhill*, Dundalk

Libby, L.M., *et al.* (1976) 'Isotopic Tree Thermometers', *Nature, 261*, 284-8

Liese, W. (1978) 'Bruno Huber; the Pioneer of European Dendrochronology', *BAR*, International series, *51*, 1-10

Lowson, J.M. (1966) *Textbook of Botany*, London

Lucas, A.T. (1953) 'The Horizontal Mill in Ireland', *J. Roy. Soc. Antiq. Ireland, 83*, 1-36

—— (1955) 'Horizontal Mill, Ballykilleen, Co. Offaly', *J. Roy. Soc. Antiq. Ireland, 85*, 100-13

Mallory, J.P., and Baillie, M.G.L. (1981) 'Tech darach', *Irish Archaeol. Research Forum*, volume in honour of Dr J. Raftery (forthcoming)

McAdam, R. (1856) 'Ancient Watermills', *Ulster J. Archaeol., 1* (4), 6-15

McCourt, D. (1966) 'Some Cruck-framed Buildings in Donegal and Derry', *Ulster Folklife, 11*, 39-50

—— and Evans, D. (1973) 'A Seventeenth Century Farmhouse at Liffock, Co. Londonderry', *Ulster J. Archaeol., III* (35), 48-56

McCracken, E. (1947) 'The Woodlands of Ulster in the Early Seventeenth Century', *Ulster J. Archaeol., III*, 15-25

—— (1971) *The Irish Woods since Tudor Times*, Newton Abbot

McGrail, S. (1978a) 'Log-boats of England and Wales with Comparative Material from European and Other Countries', *BAR*, British series, *51*, 140

—— (1978b) 'Dating Wooden Boats', *BAR,* International series, *51,* 239-58

McKerrell, M. (1975) 'Correction Procedures for C-14 Dates' in T. Watkins (ed.), *Radiocarbon: Calibration and Prehistory,* Edinburgh, pp. 47-100

McSkimin, S. (1909) *History and Antiquities of Carrickfergus,* Belfast

Milsom, S. (1979) 'Within and Between Tree Variation in Certain Properties of Annual Rings of Sessile Oaks', PhD dissertation, Liverpool Polytechnic

Morgan, R.A. (1977) 'Dendrochronological Dating of a Yorkshire Timber Building', *Vernacular Architecture, 8,* 809-14

Morgan, R.A., Coles, J.N., and Orme, B.J. (1978) 'Tree-ring Studies in the Somerset Levels', *BAR,* International series, *51,* 211-22

—— and Schofield, J. (1978) 'Tree-rings and the Archaeology of the Thames Waterfront in the City of London', *BAR,* International series, *51,* 223-38

Morris, M.G., and Perring, F.H. (1974) *The British Oak,* Faringdon

Oldrieve, W.T. (1916) 'The Ancient Roof of Glasgow Cathedral', *Proc. Soc. Antiq. Scotland, 50,* 155-73

ORiordáin, B. (1971) 'Excavations at High Street and Winetavern Street Dublin', *Medieval Archaeol., 15,* 73-85

—— (1973) 'Report on the Dublin Excavations', *Medieval Archaeol., 17,* 151-2

Otlet, R.L., and Walker, A.J. (1979) 'Harwell Radiocarbon Measurements III', *Radiocarbon, 21* (3), 358-83

Paterson, T.G.F. (1960) 'Housing and House Types in Co. Armagh', *Ulster Folklife, 6,* 7-17

Pearson, G.W. (1979) 'Precise $^{14}$C Measurement by Liquid Scintillation Counting', *Radiocarbon,* 21 (1), 1-21

—— (1980) 'High Precision Radiocarbon Dating by Liquid Scintillation Counting Applied to Radiocarbon Timescale Measurements', in M. Stuiver and R. Kra (eds.), *Radiocarbon, 22* (2), 337-45

—— , Pilcher, J.R., Baillie, M.G.L., and Hillam, J. (1977) 'Absolute Radiocarbon Dating Using a Low Altitude European Tree-ring Calibration', *Nature, 270* (5632), 25-8

Pilcher, J.R. (1973) 'Tree-ring Research in Ireland', *Tree-Ring Bulletin, 33,* 1-6

—— and Baillie, M.G.L. (1978) 'Implications of a European Radiocarbon Calibration', *Antiquity, 52,* 217-22

—— , Hillam, J., Baillie, M.G.L., and Pearson, G.W. (1977) 'A Long Sub-fossil Tree-ring Chronology from the North of Ireland', *New Phytol., 79,* 713-29

Rahtz, P., and Sheridan, K. (1971) 'Tamworth', *Current Archaeol., 3* (6), 164-8

Rigold, S.E. (1975) 'Structural Aspects of Medieval Timber Bridges', *Medieval Archaeol., 19,* 48-91

Robinson, W.J. (1976) 'Tree-ring Dating and Archaeology in the American Southwest', *Tree-Ring Bulletin, 36,* 9-20

Schove, D.J. (1974) 'Dendrochronological Dating of Oak from Old Windsor', *Medieval Archaeol., 18,* 165-72

—— (1979) 'Dark Age Tree-ring Dates AD 490-850', *Medieval Archaeol., 23,* 219-23

—— and Lowther, A.W.G. (1957) 'Tree-rings and Medieval Archaeology', *Medieval Archaeol, 1,* 78-95

Schweingruber, F.H., Braker, O.U., and Schar, E. (1979) 'Dendroclimatic Studies on Conifers from Central Europe and Great Britain', *Boreas, 8,* 427-52

Siebenlist Kerner, V. (1978) 'The Chronology for Certain Hillside Oaks from Western England and Wales, *BAR,* International series, *51,* 157-61

Smiley, T.L. (1958) 'The Geology and Dating of Sunset Crater, Flagstaff, Arizona' in *Guidebook to the Black Mesa Basin,* Sorocco, New Mexico, pp. 186-90

Smith, A.G., Pearson, G.W., and Pilcher, J.R. (1971) 'Belfast Radiocarbon Dates III', *Radiocarbon, 13,* 123
—— —— —— (1973) 'Belfast Radiocarbon Dates V', *Radiocarbon, 15,* 216
Smith, A.G., and Pilcher, J.R. (1973) 'Radiocarbon Dates and Vegetational History of the British Isles', *New Phytol., 72,* 903-14
Stallings, W.S. (1937) 'Some Early Papers on Tree-rings', *Tree-Ring Bulletin, 3,* 27-8
Stokes, G.T. (1900) *Some Worthies of the Irish Church,* Dublin
Studhalter, R.A. (1956) 'Early History of Crossdating', *Tree-Ring Bulletin, 21,* 31-5
Stuiver, M. (1978) 'Radiocarbon Timescale Tested against Magnetic and Other Dating Methods', *Nature, 273,* 271-4
Suess, H.E. (1970) 'Bristlecone Pine Calibration of the Radiocarbon Timescale 5200 BC to the Present' in I. Olsson (ed.), *Radiocarbon Variations and Absolute Chronology,* New York, pp. 303-12
Swiny, H.W., and Katzev, M.L. (1973) 'The Kyrenia Shipwreck' in D.J. Blackman (ed.), *Marine Archaeology,* London
Thompson, M.W. (1967) *Novgorod the Great,* London
Varley, G.C., and Gradwell, G.R. (1962) 'The Effect of Partial Defoliation by Caterpillars on the Timber Production of Oak Trees in England', *XI Internationaler Kongress fur Entomologie Wien 1960,* Sonderdruck aus den Verhandlungen, Bd II, pp. 211-14
Walker, F.S. (1978) 'Pedunculate and Sessile Oaks; Species Determination from their Wood', *BAR,* International series, *51,* 329-38
Wallace, P. (1976) 'Summary Account of Excavations at Wood Quay Dublin' in T.G. Delaney (ed.), *Excavations,* 31-2
Watson, G.P.H. (1922) 'The Development of Caerlaverock Castle', *Proc. Soc. Antiq. Scotland, 57,* 29-40
Zamotorin, I.M. (1959) 'Relative Chronology of the Pazyryk Burials', *Soviet Archaeol., 1,* 21-30
Zeuner, F.E. (1958) *Dating the Past – an Introduction to Geochronology,* London

# Index